D0889675

Jochen Hoefs

Stable Isotope Geochemistry

Springer

Berlin
Heidelberg
New York
Barcelona
Budapest
Hong Kong
London
Milan
Paris
Santa Clara
Singapore
Tokyo

Jochen Hoefs

Stable Isotope Geochemistry

4th, Completely Revised, Updated, and Enlarged Edition

With 73 Figures and 22 Tables

 Springer

Professor Dr. Jochen Hoefs
Institute of Geochemistry
University of Göttingen
Goldschmidtstraße 1
D-37077 Göttingen
Germany

ISBN 3-540-61126-6 Springer-Verlag Berlin Heidelberg New York

Library of Congress Cataloging-in-Publication Data
Hoefs, Jochen.
   Stable isotope geochemistry/Jochen Hoefs. – 4th. completely   rev., updated, and enl. ed.   p. cm.   Includes bibliographical references and index.
   ISBN 3-540-61126-6 (hardcover)
   1. Geochemistry.   2. Isotope   geology.   I. Title.   QE515.H54   1997
551.9–dc20   96-31362

Printed in Germany

SPIN 10510065   32/3135 – 5 4 3 2 1 0 – Printed on acid-free paper

Preface

Stable isotope geochemistry has become an essential part of geochemistry and has contributed significantly to the solution of a wide variety of geological problems, which span the whole field of earth sciences, from paleoclimatology to cosmochemistry, from oceanography to mantle geochemistry, to list only a few. In some fields such as ore deposit studies, stable isotopes have played an integral part for many years; in others such as environmental studies the application of stable isotopes is still growing. In recent years new microanalytical techniques, permitting relatively precise analysis of very small sample sizes, have opened up exciting research avenues which will allow the investigation of a new generation of problems.

These latest developments make a complete revision of the 3rd edition necessary. Although the new edition follows the subdivision of the earlier ones, it has been totally rewritten on the basis of the literature which has appeared since 1987. I have again tried to give a well-balanced discussion of the whole field, although I do not claim that every aspect has been considered and that no omission can be found. The book is not written primarily for the specialist in the field of stable isotope geochemistry, but more for the non-specialist and graduate student, who needs practical knowledge of how to interpret stable isotope ratios.

My colleagues Tony Fallick, Russ Harmon, and Antonio Longinelli have reviewed an early draft of the manuscript, which is gratefully acknowledged. Special thanks go to Russ Harmon, who has considerably improved the clarity of style and presentation. I take full responsibility, however, for any shortcomings that remain.

Göttingen, April 1996   Jochen Hoefs

Table of Contents

Chapter 3
Variations of Stable Isotope Ratios in Nature

CHAPTER 1

Theoretical and Experimental Principles

1.1
General Characteristics of Isotopes

Isotopes are atoms whose nuclei contain the same number of protons but a different number of neutrons. The term "isotope" is derived from the Greek (meaning equal places) and indicates that isotopes occupy the same position in the periodic table.

It is convenient to denote isotopes in the form $_n^m E$, where the superscript "m" denotes the mass number (i.e., sum of the number of protons and neutrons in the nucleus) and the subscript "n" denotes the atomic number of an element E. For example, $_6^{12}C$ is the isotope of carbon, which has six protons and six neutrons in its nucleus. The atomic weight of each naturally occurring element is the average of the weights contributed by its various isotopes.

Isotopes can be divided into two fundamental kinds, stable and unstable (radioactive) species. The number of stable isotopes is about 300, while over 1200 unstable ones have been discovered so far. The term "stable" is relative, depending on the detection limits of radioactive decay times. In the range of atomic numbers from 1 (H) to 83 (Bi), stable nuclides of all masses except 5 and 8 are known. Only 21 elements are pure elements, in the sense that they have only one stable isotope. All other elements are mixtures of at least two isotopes. In some elements, the less abundant isotope may be present in substantial proportions. In copper, for example, $_{29}^{63}Cu$ accounts for 69% of the total Cu nucleus and $_{29}^{65}Cu$ accounts for 31%. In most cases, however, one isotope is predominant, the others being present only in trace amounts.

The stability of nuclides is characterized by several important rules, two of which are briefly discussed here. The first is the so-called symmetry rule, which states that in a stable nuclide with a low atomic number, the number of protons is approximately equal to the number of neutrons, or the neutron-to-proton ratio, N/Z, is approximately equal to unity. In stable nuclei with more than 20 protons or neutrons, the N/Z ratio is always greater than unity, with a maximum value of about 1.5 for the heaviest stable nuclei. The electrostatic Coulomb repulsion of the positively charged protons grows rapidly with increasing Z. To maintain the stability in the nuclei, electrically more neutral neutrons than protons are incorporated into the nucleus (see Fig. 1).

The second rule is the so-called Oddo-Harkins rule, which states that nuclides of even atomic numbers are more abundant than those with odd numbers. As shown in Table 1, the most common of the four possible combinations is even-even, the least common odd-odd. The same relationship is demonstrated in Fig. 2, which shows that there are more stable isotopes with even than with odd proton numbers.

Radioactive isotopes can be classified as being either artificial or natural. Only the latter are of interest in geology, because they are the basis for radiometric age-dating methods. Radioactive decay processes are spontaneous nuclear reactions

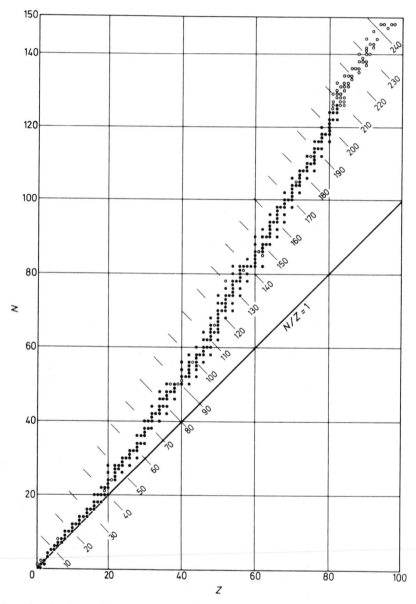

Fig. 1. Plot of number of protons (Z) and number of neutrons (N) in stable (●) and unstable (○) nuclides

Table 1. Types of atomic nuclei and their frequency of occurrence

Z – N combination	Number of stable nuclides
Even – even	160
Even – odd	56
Odd – even	50
Odd – odd	5

Fig. 2. Number of stable isotopes of elements with even and odd numbers of protons (radioactive isotopes with half-lives greater than 10^9 years are included)

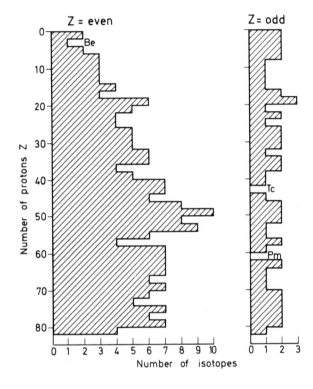

and may be characterized by the radiation emitted. Decay processes involve the emission of radiation and, in some, electrons are captured by the nucleus as well.

Radioactive decay is one process that produces isotope abundance variations. A second cause of isotope abundance differences is isotope fractionation caused by small chemical and physical differences between the isotopes of an element. It is exclusively this process important in the geochemistry of the stable isotopes that will be discussed in this book.

1.2
Isotope Effects

Differences in chemical and physical properties arising from variations in atomic mass of an element are called "isotope effects." It is well known that the electronic structure of an element essentially determines its chemical behavior, whereas the nucleus is more or less responsible for its physical properties. Because all isotopes of a given element contain the same number and arrangement of electrons, a far-reaching similarity in chemical behavior is the logical consequence. However, this similarity is not unlimited; certain differences exist in physicochemical properties due to isotope mass differences. The replacement of any atom in a molecule by one of its isotopes produces a very small perturbation in chemical behavior. However, the addition of one neutron can, for instance, depress the rate of chemical reaction considerably. Furthermore, it leads, for example, to a shift of the lines in the Raman and infrared spectra. Such mass differences are most pronounced among the lightest elements. For example, some differences in physicochemical properties of $H_2^{16}O$, $D_2^{16}O$, $H_2^{18}O$ are listed in Table 2. To summarize, the properties of molecules differing only in isotopic substitution are qualitatively the same, but quantitatively different.

Since the discovery of the isotopes of hydrogen by Urey et al. (1932a,b), differences in the chemical properties of the isotopes of the elements H, C, N, O, S, and other elements have been calculated by the methods of statistical mechanics and also determined experimentally. These differences in chemical properties can lead to considerable isotope effects in chemical reactions.

The theory of isotope effects and a related isotope fractionation mechanism will be discussed very briefly. For a more detailed introduction to the theoretical background, see Bigeleisen and Mayer (1947), Urey (1947), Melander (1960), Bigeleisen (1965), Bottinga and Javoy (1973), Javoy (1977), Richet et al. (1977), Hulston (1978), O'Neil (1986), and others.

Differences in the physicochemical properties of isotopes arise as a result of quantum mechanical effects. Figure 3 shows schematically the energy of a diatomic molecule as a function of the distance between the two atoms. According to the quantum theory, the energy of a molecule is restricted to certain discrete energy levels. The lowest level is not at the minimum of the energy curve, but above it by an amount of $^1/_2\,hv$, where h is Planck's constant and v is the frequency with which the atoms in the molecule vibrate with respect to one another. Thus, even in the ground state at a temperature of absolute zero, the vibrating molecule possesses a certain zero point energy above the minimum of the potential energy curve of the molecule. It vibrates with its fundamental frequency, which depends on the mass of the isotopes. In this context, it is important to note that only vibrational motions cause chemical isotope effects; rotational and translational motions have no effect on isotope separations. Therefore, different

Table 2. Characteristic physical properties of H_2O, D_2O, and $H_2^{18}O$

Property	$H_2^{16}O$	$D_2^{16}O$	$H_2^{18}O$
Density (20 °C, in g cm^{-3})	0.997	1.1051	1.1106
Temperature of greatest density (°C)	3.98	11.24	4.30
Melting point (760 Torr, in °C)	0.00	3.81	0.28
Boiling point (760 Torr, in °C)	100.00	101.42	100.14
Vapor pressure (at 100 °C, in Torr)	760.00	721.60	
Viscosity (at 20 °C, in centipoise)	1.002	1.247	1.056

Fig. 3. Schematic potential-energy curve for the interaction of two atoms in a stable molecule or between two molecules in a liquid or solid. (After Bigeleisen 1965)

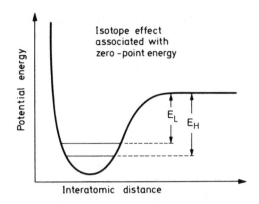

isotopic species will have different zero-point energies in molecules with the same chemical formula: the molecule of the heavy isotope will have a lower zero-point energy than the molecule of the light isotope. This is shown schematically in Fig. 3, where the upper horizontal line (E_L) represents the dissociation energy of the light molecule and the lower line (E_H) that of the heavy one. E_L is actually not a line, but an energy interval between the zero-point energy level and the "continuous" level. This means that the bonds formed by the light isotope are weaker than bonds involving the heavy isotope. Thus, during a chemical reaction, molecules bearing the light isotope will, in general, react slightly more readily than those with the heavy isotope.

1.3
Isotope Fractionation Processes

The partitioning of isotopes between two substances or two phases of the same substance with different isotope ratios is called "isotope fractionation." The main phenomena producing isotope fractionations are:

    1. Isotope exchange reactions.

    2. Kinetic processes, which depend primarily on differences in reaction rates of isotopic molecules.

1.3.1
Isotope Exchange

Isotope exchange includes processes with very different physico-chemical mechanisms. Here, the term "isotope exchange" is used for all situations in which there is no net reaction, but in which the isotope distribution changes between different chemical substances, between different phases, or between individual molecules.

    Isotope exchange reactions are a special case of general chemical equilibrium and can be written:

$$aA_1 + bB_2 = aA_2 + bB_1 \tag{1}$$

where the subscripts indicate that species A and B contain either the light or heavy isotope 1 or 2, respectively. For this reaction the equilibrium constant is expressed by:

$$K = \frac{\left(\dfrac{A_2}{A_1}\right)^a}{\left(\dfrac{B_2}{B_1}\right)^b} \tag{2}$$

where the terms in parentheses may be, for example, the molar ratios of any species. Using the methods of statistical mechanics, the isotopic equilibrium constant may be expressed in terms of the partition functions Q of the various species:

$$K = \frac{Q_{A_2}}{Q_{A_1}} \bigg/ \frac{Q_{B_2}}{Q_{B_1}} \tag{3}$$

Thus, the equilibrium constant then is simply the product or quotient of two partition function ratios, one for the two isotopic species of A, the other for B. The partition function is defined by:

$$Q = \Sigma_i(g_i \exp\,(-E_i/kT) \tag{4}$$

where the summation is over all the allowed energy levels, E_i, of the molecules and g_i is the degeneracy or statistical weight of the ith level (of E_i), k is the Boltzmann constant, and T is the temperature. Urey (1947) has shown that for the purpose of calculating partition function ratios of isotopic molecules, it is very convenient to introduce, for any chemical species, the ratio of its partition function to that of the corresponding isolated atom, which is called the reduced partition function. This reduced partition function ratio can be manipulated in exactly the same way as the normal partition function ratio. The partition function of a molecule can be separated into factors corresponding to each type of energy: translation, rotation, and vibration:

$$Q_2/Q_1 = (Q_2/Q_1)_{trans} \times (Q_2/Q_1)_{rot} \times (Q_2/Q_1)_{vib} \tag{5}$$

The difference of the translation and rotation energy is more or less the same among the compounds appearing at the left- and right-hand side of the exchange reaction equation, except for hydrogen, where rotation must be taken into account. This leaves differences in vibrational energy as the predominant source of "isotope effects." The vibrational energy term can be separated into two components. The first is related to the zero-point energy difference and accounts for most of the variation with temperature. The second term represents the contributions of all the other bound states and is not very different from unity. The complications which may occur relative to this simple model are mainly that the oscillator is not perfectly harmonic, so an "anharmonic" correction has to be added.

For geological purposes the dependence of the equilibrium constant K on temperature is the most important property (Eq. 3). In principle, isotope fractionation factors for isotope exchange reactions are also slightly pressure dependent because isotopic substitution makes a minute change in the molar volume of solids and liquids. Experimental studies of pressures up to 20 kbar by Clayton et al. (1975) have shown that the pressure dependence for oxygen is, however, less than the limit of analytical detection. Thus, as far as it is known today, the pressure dependence seems to be of no importance for crustal and upper mantle environments (but see Polyakov and Kharlashina 1994).

No isotope fractionation occurs at very high temperatures. However, isotope fractionations do not decrease to zero monotonically with increasing temperatures. At higher temperatures, fractionations may change sign (called crossover) and may increase in magnitude, but they must approach to zero at very high temperatures. Such crossover phenomena are due to the complex manner by which thermal excitation of the vibration of atoms contributes to an isotope effect (Stern et al. 1968).

For ideal gas reactions, there are two temperature regions where the behavior of the equilibrium constant is simple: at low temperatures (generally much below room temperature) K follows $\ln K \sim 1/T$, where T is the absolute temperature, and at high temperatures the approximation becomes $\ln K \sim 1/T^2$.

The temperature ranges at which these simple behaviors are exhibited depend on the vibrational frequencies of the molecules involved in the reaction. We have seen that for the calculation of a partition function ratio for a pair of isotopic molecules, the vibrational frequencies of each molecule must be known. When solid materials are considered, the evaluation of partition function ratios becomes even more complicated, because it is necessary to consider not only the independent internal vibrations of each molecule, but also the lattice vibrations.

1.3.1.1
Fractionation Factor (α)

Usually, the fractionation factor (α) is of more interest than the equilibrium constant. The fractionation factor is defined as the ratio of the numbers of any two isotopes in one chemical compound A divided by the corresponding ratio for another chemical compound B:

$$\alpha_{A-B} = \frac{R_A}{R_B} \qquad (6)$$

If the isotopes are randomly distributed over all possible positions in the compounds A and B, then α is related to the equilibrium constant K by:

$$\alpha = K^{1/n} \qquad (7)$$

where n is the number of atoms exchanged. For simplicity, isotope exchange reactions are written such that only one atom is exchanged (Eq. 7). In these cases, the equilibrium constant is identical to the fractionation factor. For example, the fractionation factor for the exchange of ^{18}O and ^{16}O between water and $CaCO_3$ is expressed as follows:

$$H_2{}^{18}O + \frac{1}{3}CaC^{16}O_3 \Leftrightarrow H_2{}^{16}O + \frac{1}{3}CaC^{18}O_3 \tag{8}$$

with the fractionation factor $\alpha CaCO_3\text{-}H_2O$ defined as:

$$\alpha_{CaCO_3-H_2O} = \frac{\left(\dfrac{^{18}O}{^{16}O}\right)_{CaCO_3}}{\left(\dfrac{^{18}O}{^{16}O}\right)_{H_2O}} = 1.031 \text{ at } 25\ ^\circ C \tag{9}$$

1.3.1.2
The Delta Value (δ)

In isotope geochemistry, it is common practice to express isotopic composition in terms of "delta" (δ) values. For two compounds A and B whose isotopic composition has been measured in the laboratory by conventional mass spectrometry:

$$\delta_A = \left(\frac{R_A}{R_{St}} - 1\right) \cdot 10^3\ (\permil) \tag{10}$$

and

$$\delta_B = \left(\frac{R_B}{R_{St}} - 1\right) \cdot 10^3\ (\permil) \tag{11}$$

where R_A and R_B are the respective isotope ratio measurements for the two compounds and R_{St} is the defined isotope ratio of a standard sample.

For the two compounds A and B, the δ-values and fractionation factor α are related by:

$$\delta_A - \delta_B = \Delta_{A\text{-}B} \approx 10^3 \ln \alpha_{A\text{-}B} \tag{12}$$

Table 3 illustrates the closeness of the approximation. Considering the experimental uncertainties in isotope ratio determinations, these approximations are excellent for differences in δ-values of less than about 10.

Table 3. Comparison between δ, α, and $10^3 \ln\alpha_{A\text{-}B}$

δ_A	δ_B	$\Delta_{A\text{-}B}$	$\alpha_{A\text{-}B}$	$10^3 \ln\alpha_{A\text{-}B}$
1.00	0	1	1.001	0.9995
10.00	0	10	1.01	9.95
20.00	0	20	1.02	19.80
10.00	5.00	4.98	1.00498	4.96
20.00	15.00	4.93	1.00493	4.91
30.00	20.00	9.80	1.00980	9.76
30.00	10.00	19.80	1.01980	19.61

1.3.1.3
Evaporation-Condensation Processes

Of special interest in stable isotope geochemistry are evaporation-condensation processes, because differences in the vapor pressures of isotopic compounds lead to significant isotope fractionations. For example, from the vapor pressure data for water given in Table 2, it is evident that the lighter molecular species are preferentially enriched in the vapor phase, the extent depending upon the temperature. Such an isotopic separation process can be treated theoretically in terms of fractional distillation or condensation under equilibrium conditions as is expressed by the Rayleigh (1896) equation. For a condensation process this equation is:

$$\frac{R_v}{R_{v_o}} = f^{a-1} \tag{13}$$

where R_{v_o} is the isotope ratio of the initial bulk composition and R_v is the instantaneous ratio of the remaining vapor (v), f is the fraction of the residual vapor, and the fractionation factor α is given by R_l/R_v (l = liquid). Similarly, the instantaneous isotope ratio of the condensate leaving the vapor (R_l) is given by:

$$\frac{R_l}{R_{v_o}} = \alpha f^{\alpha-1} \tag{14}$$

and the average isotope ratio of the separated and accumulated condensate (R_l) at any time of condensation is expressed by:

$$\frac{\bar{R}_l}{R_{v_o}} = \frac{1-f^\alpha}{1-f} \tag{15}$$

For a distillation process the instantaneous isotope ratios of the remaining liquid and the vapor leaving the liquid are given by:

$$\frac{R_l}{R_{l_o}} = f^{\left(\frac{1}{\alpha}-1\right)} \tag{16}$$

and

$$\frac{R_v}{R_{l_o}} = \frac{1}{\alpha} f^{\left(\frac{1}{\alpha}-1\right)} \tag{17}$$

The average isotope ratio of the separated and accumulated vapor is expressed by:

$$\frac{\bar{R}_v}{R_{l_o}} = \frac{1-f^{\frac{1}{\alpha}}}{1-f} \quad (f = \text{fraction of residual liquid}) \tag{18}$$

Any isotope fractionation occurring in such a way that the products are isolated from the reactants immediately after formation will show a characteristic

Fig. 4. $\delta^{18}O$ in a cloud vapor and condensate plotted as a function of the fraction of remaining vapor in the cloud for a Rayleigh process. The temperature of the cloud is shown *on the lower axis.* The increase in fractionation with decreasing temperature is taken into account. (After Dansgaard 1964)

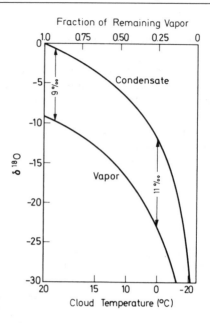

trend in isotopic composition. As condensation or distillation proceed, the residual vapor or liquid will become progressively depleted or enriched with respect to the heavy isotope. A natural example is the fractionation between oxygen isotopes in the water vapor of a cloud and the raindrops released from the cloud. The resulting decrease of the $^{18}O/^{16}O$ ratio in the residual vapor and the instantaneous isotopic composition of the raindrops released from the cloud are shown in Fig. 4 as a function of the fraction of vapor remaining in the cloud.

1.3.2
Kinetic Effects

The second main group of phenomena producing fractionations are kinetic isotope effects, which are associated with incomplete and unidirectional processes such as evaporation, dissociation reactions, biologically mediated reactions, and diffusion. The latter process is of special significance for geological purposes, which warrants separate treatment (Sect. 1.3.3). A kinetic isotope effect also occurs when the rate of a chemical reaction is sensitive to atomic mass at a particular position in one of the reacting species.

The theory of kinetic isotope fractionations has been discussed by Bigeleisen and Wolfsberg (1958), Melander (1960), and Melander and Saunders (1980). Knowledge of kinetic isotope effects is very important, because it can provide information about details of reaction pathways.

Quantitatively, many observed deviations from simple equilibrium processes can be interpreted as consequences of the various isotopic components having different rates of reaction. Isotope measurements taken during unidirectional chemical reactions always show a preferential enrichment of the lighter isotope in the reaction products. The isotope fractionation introduced during the course of a unidirectional reaction may be considered in terms of the ratio of rate constants for the isotopic substances. Thus, for two competing isotopic reactions:

$$A_1 \xrightarrow{k_1} B_1, \text{ and } A_2 \xrightarrow{k_2} B_2 \tag{19}$$

the ratio of rate constants for the reaction of light and heavy isotope species k_1/k_2, as in the case of equilibrium constants, is expressed in terms of two partition function ratios, one for the two reactant isotopic species, and one for the two isotopic species of the activated complex or transition state A^*:

$$\frac{k_1}{k_2} = \left[\frac{Q^*_{(A_2)}}{Q^*_{(A_1)}} \middle/ \frac{Q^*_{(A_2)}}{Q^*_{(A_1)}} \right] \frac{v_1}{v_2} \tag{20}$$

The factor v_1/v_2 in the expression is a mass term ratio for the two isotopic species. The determination of the ratio of rate constants is, therefore, principally the same as the determination of an equilibrium constant, although the calculations are not so precise because of the need for detailed knowledge of the transition state. By "transition state" is meant that molecular configuration which is most difficult to attain along the path between the reactants and the products. This theory is based on the idea that a chemical reaction proceeds from some initial state to a final configuration by a continuous change, and that there is some critical intermediate configuration called the activated species or transition state. There are a small number of activated molecules in equilibrium with the reacting species and the rate of reaction is controlled by the rate of decomposition of these activated species.

1.3.3
Diffusion

Diffusion is a major pathway for material transport in a wide variety of geological environments. At elevated temperatures, diffusion will be the rate-controlling step of mass transport in the absence of fluid advection or mineral dissolution/precipitation.

The process of diffusion can cause significant isotope fractionations. In general, light isotopes are more mobile and therefore more affected by diffusion than heavy isotopes. For gases, the ratio of diffusion coefficients is equivalent to the square root of their masses. Consider the isotopic molecules of carbon in CO_2, with masses $^{12}C^{16}O^{16}O$ and $^{13}C^{16}O^{16}O$ having molecular weights of 44 and 45. Solving the expression equating the kinetic energies ($^1/_2\, m\, v^2$) of both species, the ratio of velocities is equivalent to the square root of 45/44 or 1.01. That is, regardless of temperature, the average velocity of $^{12}C^{16}O^{16}O$ molecules is about 1% greater than the average velocity of $^{13}C^{16}O^{16}O$ molecules in the same system. This isotope effect, however, is more or less limited to ideal gases, where collisions between molecules are infrequent and intermolecular forces negligible. Under geological conditions,

gases are characterized by higher pressures and frequent gas collisions lead to isotope effects which are considerably smaller than theoretically calculated.

In solutions and solids the relationships are much more complicated. The term
"solid-state diffusion" generally includes volume diffusion and diffusion mechanisms, where the atoms move along paths of easy diffusion such as grain boundaries and surfaces. Diffusive-penetration experiments indicate a marked enhancement of diffusion rates along grain boundaries which are orders of magnitude faster than for volume diffusion. Thus, grain boundaries can act as pathways
of rapid exchange. Volume diffusion is driven by the random thermal motion of
an element or isotope within a crystal lattice, depending on the presence of point
defects, such as vacancies or interstitial atoms, within the lattice.

The flux F of elements or isotopes diffusing through a medium is proportional to the concentration gradient (dc/dx) such that:

$$F = -D \, (dc/dx) \text{ (Fick's first law)} \tag{21}$$

where D represents the diffusion coefficient, and the minus sign denotes that the
concentration gradient has a negative slope, i.e., elements or isotopes move from
points of high concentration towards points of low concentration. The diffusion
coefficient D varies with temperature according to the Arrhenius relation:

$$D = D_0^{(-Ea/RT)} \tag{22}$$

where D_0 is the temperature-independent factor, Ea is the activation energy, and
R is the gas constant.

In recent years there have been several attempts to determine diffusion coefficients, mostly utilizing secondary ion mass spectrometry (SIMS), where isotope
compositions have been measured as a function of depth below a crystal surface.

Fig. 5. Arrhenius plot of
diffusion coefficients versus
reciprocal temperatures for
various minerals.
(After Eiler et al. 1992)

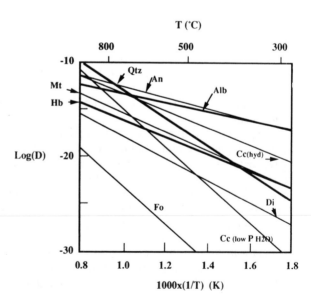

Fig. 6. Oxygen isotope
changes of quartz, feldspar,
and hornblende in a rock undergoing diffusive exchange
during cooling.
(After Eiler et al. 1992)

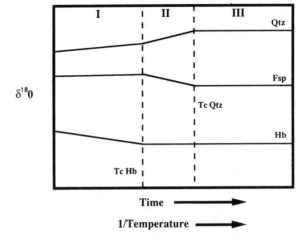

A plot of the logarithm of the diffusion coefficient versus reciprocal temperature yields a linear relationship over a significant range of temperature for most
minerals. Such an Arrhenius plot for various minerals is shown in Fig. 5, which illustrates the variability in diffusion coefficients for different minerals. The practical application of this fact is that the different minerals in a rock will exchange
oxygen at different rates and become closed systems to isotopic exchange at different temperatures. As a rock cools from the peak of a thermal event, isotope
equilibrium fractionations between the minerals will increase. The rate at which
the coexisting minerals can approach the lower temperature equilibrium is limited by the volume diffusion rates of the respective minerals.

Several models for diffusive transport in and among minerals have been discussed in the literature, the most recent one being the "fast grain boundary (FGB)
model" of Eiler et al. (1993). The FGB model considers the effects of diffusion between nonadjacent grains and shows that, when mass balance terms are included,
closure temperatures become a strong function of both the modal abundances of
constituent minerals and the differences in diffusion coefficients between all coexisting minerals.

Figure 6 illustrates schematically how the oxygen isotope composition of
quartz, feldspar, and hornblende changes during cooling. Early in the cooling history (period I) all three minerals exchange oxygen isotopes, with equilibrium
maintained. Later (period II) isotope exchange with hornblende has ceased and,
therefore, only quartz and feldspar continue to undergo exchange. During this period, the feldspar–hornblende ^{18}O-fractionation decreases, although no feldspar–hornblende exchange takes place. In period III, after quartz "closes," no further isotope exchange takes place and it is this isotopic state that is preserved in
the cooled rocks. Thus, the measured O-isotope fractionations between the coexisting mineral pairs are not an accurate recorder of the closure temperature. All
other mineral pairs will be affected by the closure temperature of the first mineral in the system to close, the extent of which will be a function of the modal abundance of the minerals that continue to be involved in isotopic exchange.

1.3.4
Other Factors Influencing Isotopic Fractionations

1.3.4.1
Chemical Composition

Qualitatively, the isotopic composition of a mineral depends to a very high degree upon the nature of the chemical bonds within the mineral and to a smaller degree upon the atomic mass of the respective elements. In general, bonds to ions with a high ionic potential and low atomic mass are associated with high vibrational frequencies and have a tendency to incorporate preferentially the heavy isotope. This relationship can be demonstrated by considering the bonding of oxygen to the small highly charged Si^{4+} ion compared to the relatively large Fe^{2+} ion. In natural mineral assemblages, quartz is always the most ^{18}O-rich mineral and magnetite is always the most ^{18}O-deficient mineral. Furthermore, carbonates are always enriched in ^{18}O relative to most other mineral groups because oxygen is bonded to the small, highly charged C^{4+} ion. The mass of the divalent cation is of secondary importance to the C-O bonding. However, the mass effects are apparent in ^{34}S distributions among sulfides, where, for example, ZnS always concentrates ^{34}S relative to coexisting PbS.

1.3.4.2
Crystal Structure

Structural effects are secondary in importance to those arising from the primary chemical bonding, the heavy isotope being concentrated in the more closely packed or well-ordered structures. The ^{18}O and D fractionations between ice and liquid water arise mainly from differences in the degree of hydrogen bonding (order). A relatively large isotope effect associated with structure is observed between graphite and diamond (Bottinga 1969b). With a modified increment method, Zheng (1993a) has calculated this structural effect for the SiO_2 and Al_2SiO_5 polymorphs and demonstrated that ^{18}O will be enriched in the high-pressure forms.

1.3.4.3
Non-Mass-Dependent Isotope Effects

It has been a common belief that chemically produced isotope effects arise solely because of isotopic mass differences. This means that for an element with more than two isotopes, such as oxygen or sulfur, the enrichment of ^{18}O relative to ^{16}O or ^{34}S relative to ^{32}S is expected to be approximately twice as large as the enrichment of ^{17}O relative to ^{16}O or as the enrichment of ^{33}S relative to ^{32}S. This yields a slope of about 0.5 on a three-isotope correlation diagram. On this basis, non-mass-dependent isotope fractionations have been ascribed solely to nuclear processes.

As was first demonstrated by Thiemens and Heidenreich (1983), an unusual isotope fractionation occurs when ozone is produced from an electrical discharge in pure O_2. An equal enrichment of ^{17}O and ^{18}O was observed in the product ozone, rather than $\delta^{17}O = 0.5 \, \delta^{18}O$, as expected for a mass-dependent fractionation process. Subsequently, it was demonstrated experimentally that the anomalous fractionation was associated with the actual O_3 formation process rather than the O_2 dissociation step (Heidenreich and Thiemens 1985). These authors initially suggested that self-shielding and/or molecular symmetry of O_2 might be responsible for the resultant isotopic composition of ozone. As was later shown, self-shielding can be ruled out, with symmetry factors seeming to play the decisive role in this unusual fractionation.

Mass-independent isotopic fractionations, due to symmetry factors, have also been observed in the reaction O + CO (Bhattacharya and Thiemens 1989) and in the isotopes of sulfur for the $SF_5 + SF_5$ reaction (Bains-Sahota and Thiemens 1989). The most extensively investigated example in the natural environment is the substantial enhancement in the heavy isotopes of stratospheric ozone (see Sect. 3.9).

1.3.5
Isotope Geothermometers

Isotope thermometry has become well established since the classic paper of Harold Urey (1947) about the thermodynamic properties of isotopic substances. The partitioning of two stable isotopes of an element between two mineral phases can be viewed as a special case of element partitioning between two minerals. The most important difference between the two exchange reactions is the pressure insensitivity of isotope partitioning, which represents a considerable advantage relative to the numerous types of other geothermometers, all of which exhibit a pressure dependence.

The necessary condition to apply an isotope geothermometer is isotope equilibrium. Isotope exchange equilibrium should be established during reactions whose products are in chemical and mineralogical equilibrium. Demonstration that the minerals in a rock are in oxygen isotope equilibrium is strong evidence that the rock is in chemical equilibrium. To break Al–O and Si–O bonds and allow rearrangement towards equilibrium needs sufficient energy to effect chemical equilibrium.

Theoretical studies show that the fractionation factor α for isotope exchange between minerals is a linear function of $1/T^2$, where T is temperature in degrees Kelvin. Bottinga and Javoy (1973) demonstrated that O-isotopic fractionation between anhydrous mineral pairs at temperatures >500 °C can be expressed in terms of a relationship of the form:

$$1000 \ln \alpha = A/T^2 \tag{23}$$

which means that the factor A has to be known in order to calculate a temperature of equilibration. By contrast, fractionations at temperatures <500 °C can be expressed by an equation of the form:

$$1000 \ln \alpha = A/T^2 + B \tag{24}$$

One drawback to isotope thermometry in slowly cooled metamorphic and magmatic rocks is that temperature estimates are often significantly lower than those from other geothermometers. This results from isotopic resetting associated with retrograde isotope exchange between coexisting phases or with transient fluids. During cooling in closed systems, volume diffusion may be the principal mechanism by which isotope exchange occurs between coexisting minerals.

Giletti (1986) proposed a model in which experimentally derived diffusion data can be used in conjunction with measured isotope ratios to explain disequilibrium isotope fractionations in slowly cooled, closed-system mineral assemblages. This approach describes diffusional exchange between a mineral and an infinite reservoir whose bulk isotopic composition is constant during exchange. However, mass balance requires that loss or gain of an isotope from one mineral must be balanced by a change in the other minerals still subject to isotopic exchange. Recent numerical modeling by Eiler et al. (1992) has shown that closed-system exchange depends not only on modal proportions of all of the minerals in a rock, but also mineral diffusivities, grain size, grain shape, and cooling rate. Further complications may arise in the presence of fluids, because isotope exchange may also occur by solution-reprecipitation or chemical reaction rather than solely by diffusion.

Three different methods have been used to determine the equilibrium constants for isotope exchange reactions:
1. Theoretical calculations
2. Experimental determinations in the laboratory
3. Calibration on an empirical basis

Method 3 is based on the idea that the calculated "formation temperature" of a rock (calculated from other geothermometers) serves as a calibration to the measured isotopic fractionations, assuming that all minerals were at equilibrium. However, because there is evidence that totally equilibrated systems are not very common in nature, such empirical calibrations should be regarded with caution.

1.3.5.1
Theoretical Calculations

Calculations of equilibrium isotope fractionation factors have been particularly successful for gases. Richet et al. (1977) calculated the partition function ratios for a large number of gaseous molecules. They demonstrated that the main source of error in the calculation is the uncertainty in the vibrational molecular constants.

The theory developed for perfect gases could be extended to solids if the partition functions of crystals could be expressed in terms of a set of vibrational frequencies that correspond to its various fundamental modes of vibration (O'Neil 1986). By estimating thermodynamic properties from elastic, structural, and spectroscopic data, Kieffer (1982) calculated oxygen isotope partition function ratios and from these calculations derived a set of fractionation factors for silicate minerals. The calculations have no inherent temperature limitations and can be applied to any phase for which adequate spectroscopic and mechanical data are available. They are, however, limited in accuracy as a consequence of the approximations needed to carry out the calculations and the limited accuracy of the spectroscopic data.

Isotope fractionations in solids depend on the nature of the bonds between atoms of an element and the nearest atoms in the crystal structure (O'Neil 1986). The correlation between bond strength and oxygen isotope fractionation was investigated by Schütze (1980), who developed an "increment" method for predicting oxygen isotope fractionations in silicate minerals. Richter and Hoernes (1988) applied this method to the calculation of oxygen isotope fractionations between silicate minerals and water. More recently, Zheng (1991, 1993b, c) extended the increment method by using parameters of crystal chemistry with no empirical factor. Thus, the fractionation factors calculated over the temperature range 0–1200 °C are in surprisingly good agreement with experimental calibrations.

1.3.5.2
Experimental Calibrations

In general, experimental calibrations of isotope geothermometers have been performed between 250 and 800 °C. The upper temperature limit is usually determined by the stability of the mineral being studied or by limitations of the experimental apparatus, whereas the lower temperature limit is determined by the decreasing rate of exchange.

Most of the published data on mineral fractionations have been determined by exchange of single minerals with water. This approach is limited by two factors: (1) many minerals are unstable, melt, or dissolve in the presence of water and (2) the temperature dependence of the fractionation factor for aqueous systems is complicated as a consequence of the high vibrational frequencies of the water molecule. An alternative approach to the experimental determination of isotope fractionation between minerals has been employed by Clayton et al. (1989) and Chiba et al. (1989), who demonstrated that both limitations can be avoided by using $CaCO_3$, instead of H_2O, as the common exchange medium. These studies showed that most common silicates undergo rapid oxygen isotope exchange with $CaCO_3$ at temperatures above 600 °C and pressures of 15 kilobars.

Various experimental approaches have been used to determine fractionation factors. The three most common techniques are described below.

1. Two-direction approach: This method is analogous to reversing reactions in experimental petrology and is the only method by which the attainment of equilibrium can be convincingly demonstrated. Equilibrium fractionations are achieved by starting on opposite sides of the equilibrium distribution.

2. Partial-exchange technique: The partial exchange technique is used when rates of isotopic exchange are relatively low and is based on the assumption that the rates of isotope exchange for companion exchange experiments are identical. Experimental runs have to be the same in every respect except in the isotopic compositions of the starting materials. Rates of isotope exchange reactions in heterogeneous systems are relatively high at first (surface control) and then become progressively lower with time (diffusion control). Four sets of experiments are shown in Fig. 7 for the CO_2–graphite system (after Scheele and Hoefs 1992). Northrop and

Clayton (1966) presented a set of equations to describe the kinetics of isotope exchange reactions and developed a general equation for the partial exchange technique. At low degrees of exchange the fractionations determined by the partial exchange technique are often larger than the equilibrium fractionations (O'Neil 1986).

3. Three-isotope method: This method, introduced by Matsuhisa et al. (1979) and later modified by Matthews et al. (1983a), uses the measurement of both $^{17}O/^{16}O$ and $^{18}O/^{16}O$ fractionations in a single experiment that has gone to equilibrium. The initial $^{18}O/^{16}O$ fractionation for the mineral–fluid system is selected to be close to the assumed equilibrium, while the initial $^{17}O/^{16}O$ fractionation is chosen to be very different from the equilibrium value. In this way the change in the $^{17}O/^{16}O$ fractionations monitors the extent of isotopic exchange and the $^{18}O/^{16}O$ fractionations reflect the equilibrium value. Figure 8 is a schematic diagram of the three-isotope exchange method.

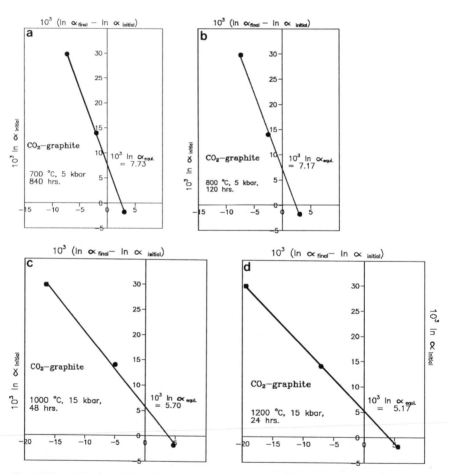

Fig. 7. CO_2-graphite partial-exchange experiments in a Northrop and Clayton plot at 700, 800, 1000, and 1200 °C. The connecting line in experiments at 1200 °C has a plane slope and defines the intercept more precisely than the experiment at 700 °C. (After Scheele and Hoefs 1992)

Fig. 8. Schematic diagram of the three-isotope exchange method. Natural samples plotted on the primary mass fractionation line (*PF*). Initial isotopic compositions are mineral (M_o) and water (W_o), which is well removed from equilibrium with M_o in $\delta^{17}O$, but very close to equilibrium with M_o in $\delta^{18}O$. Complete isotopic equilibrium is defined by a secondary mass fractionation line (*SF*) parallel to *PF* and passing through the bulk isotopic composition of the mineral plus water system. Isotopic compositions of partially equilibrated samples are M_f and W_f and completely equilibrated samples are M_e and W_e. Values for M_e and W_e can be determined by extrapolation from the measured values of M_o, M_f, W_o, and W_f. (After Matthews et al. 1983a; Sheppard 1984)

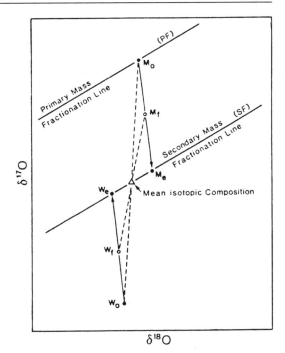

1.4
Basic Principles of Mass Spectrometry

Mass spectrometric methods are by far the most effective means of measuring isotope abundances. A mass spectrometer separates charged atoms and molecules on the basis of their masses based on their motions in magnetic and/or electrical fields. The design and the applications of the many types of mass spectrometers are too broad to cover here. Therefore, only the principles of mass analysis will be briefly discussed.

In principle, a mass spectrometer may be divided into four different central constituent parts: (1) the inlet system, (2) the ion source, (3) the mass analyzer, and (4) the ion detector (see Fig. 9).

1. Special arrangements for the *inlet system* are necessary because the instability of the ions produced and the mass separation require a high vacuum. If the mean free path length (flight without collision with other molecules) of molecules is large compared with the dimensions of the tubing through which the gas is flowing, this condition is referred to as molecular flow. During molecular flow the gas particles do not influence each other. Therefore, the gas flow velocity of the isotopically lighter component is greater than that of the isotopically heavier component, with the result that the heavier isotope becomes enriched in the reservoir from which the gas flows into the mass spectrometer. To avoid such a mass discrimination, the isotope abundance measurements of gaseous substances normally are carried out utilizing viscous gas flow. During the viscous gas flow the free path length of molecules is

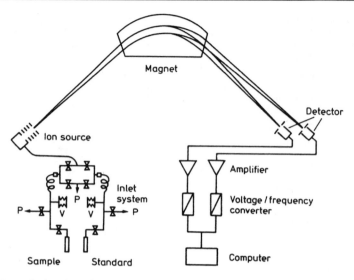

Fig. 9. Schematic drawing of a mass spectrometer for stable isotope measurements. *P* denotes pumping system, *V* denotes a variable volume

small, molecule collisions are frequent (causing the gas to be well mixed), and no mass separation takes place. The normal gas pressure is around 100 Torr. At the end of the viscous-flow inlet system, there is a "leak," a constriction in the flow line.

2. The *ion source* is the part of the mass spectrometer where ions are formed, accelerated, and focused into a narrow beam. In the ion source, the gas flow is always molecular. In general, ions are produced thermally or by electron impact. Ions of gaseous samples are most reliably produced by electron bombardment. A beam of electrons is emitted by a heated filament, usually tungsten or rhenium, and is directed to pass between two parallel plates. The beam is collimated by means of a weak magnetic field. Positive ions are formed between the two parallel plates as a result of gas molecule-electron collisions. The ions are drawn out of the electron beam by the action of an electric field, and are subsequently further accelerated up to several kilovolts. Thus, the positive ions entering the magnetic field are essentially monoenergetic, i.e., they will possess the same kinetic energy, according to the equation:

$$^{1}/_{2} Mv^2 = eV \qquad\qquad (25)$$

There is a minimum threshold energy below which ionization does not occur. The energy of electrons used for ionization generally is about 50–70 V, because this range of energy maximizes the efficiency of single ionization, but is too low to produce a significant number of multiply charged ions. The principal advantage of an electron-bombardment ion source is the stability of the resulting ion beam, the disadvantage being that the vacuum system must be extremely clean, because the electrons will ionize any gas that is present in the ionization chamber.

3. The *mass analyzer* separates the ion beams emerging from the ion source according to their m/e (mass/charge) ratios. From the many possible mass analyzer configurations, only the first-order, direction-focusing mass analyzer is used in stable isotope research. As the ion beam passes through the magnetic field, the ions are deflected into circular paths, the radii of which are proportional to the square root of m/e. Thus, the ions are separated into beams, each characterized by a particular value of m/e. In 1940, Nier introduced the sector magnetic analyzer. In this type of analyzer, deflection takes place in a wedge-shaped magnetic field. The ion beam enters and leaves the field at right angles to the boundary, so the deflection angle is equal to the wedge angle, for instance, 60°. The sector instrument has the advantage of its source and detector being comparatively free from the mass-discriminating influence of the analyzer field.

4. After passing through the magnetic field, the separated ions are collected in the *ion detector* and converted into an electrical impulse, which is then fed into an amplifier. For relatively large ion currents a simple metal cup (Faraday cage) is used. The cup is grounded through a high ohmic resistor. As the ion current passes to the ground, the potential drop in the resistor acts as a measure of the ion current.

By collecting two ions beams of the isotopes in question simultaneously, and by measuring the ratio of this ion current directly, a much higher precision can be obtained than from a single ion beam collection. With simultaneous collection, the isotope ratios of two samples can be compared quickly under nearly identical conditions. Nier et al. (1947) developed this technique for routine isotope ratio measurements and McKinney et al. (1950) improved this type of mass spectrometer, which has become the standard for isotope ratio analysis. The double-collecting mass spectrometer employed a precision voltage divider in a null circuit (Kelvin-bridge type). With this technique, the isotope ratios could be accurately measured using the chart recorder output of a vibrating reed electrometer.

During the 1960s and early 1970s, instrument makers automated their mass spectrometers, changing the measurement system from the null technique to one employing voltage-to-frequency converters and counters on each electrometer output. The most recent generation of mass spectrometers is fully automated and computerized, improving the reproducibility to values better than ±0.02‰.

The overall instrumental error of the mass-spectrometric measurement may be increased by nonlinearities within the individual measurement devices. The probable variation between different instruments may reach a level of 1%–2% of the measured δ-values. This is not critical for small differences in isotopic composition. However, the uncertainty in comparing data from different laboratories increases when samples of very different isotopic compositions are compared. Blattner and Hulston (1978), by distributing a pair of calcite reference samples, showed that the individual differences between $\delta^{18}O$ determinations by more than ten laboratories range from 23.0 to 23.6‰.

Special efforts have been undertaken in the past few years to reduce the sample size for isotope measurements. The most successful approach is the "static" measurement technique. Unlike the "dynamic" technique in which the sample gas to be analyzed is stored in a container and continuously introduced into the ion source, the "static" technique involves the introduction of the whole sample into the ion source at once, the isotope composition is measured, and the sample is then pumped away after completion of the measurement. The mass spectrometer is recalibrated at regular short intervals by introducing a standard gas of the same chemical form as the sample. With this technique subnanogram quantities can be measured with a precision better than 0.5‰ (see also Sect. 1.7).

1.5
Standards

The accuracy with which *absolute* isotope abundances can be measured is substantially poorer than the precision with which *relative* differences in isotope abundances between two samples can be determined. Nevertheless, the determination of absolute isotope ratios is very important, because these numbers form the basis for the calculation of the relative differences, the δ-values. Table 4 summarizes absolute isotope ratios of primary standards used by the international stable isotope community.

Irregularities and problems concerning standards have been evaluated by Friedman and O'Neil (1977), Gonfiantini (1978, 1984), and Coplen et al. (1983). The accepted unit of isotope ratio measurements is the delta value (δ) given in per mill (‰). The δ-value is defined as:

$$\delta \text{ in } \permil = \frac{R_{(Sample)} - R_{(Standard)}}{R_{(Standard)}} \cdot 1000 \tag{26}$$

where R represents the measured isotope ratio. If $\delta_A > \delta_B$, it is convenient to speak of A being enriched in the rare isotope or "heavier" than B. Unfortunately, not all of the δ-values cited in the literature are given relative to a single universal standard, so that often several standards of one element are in use. To convert δ-values from one standard to another, the following equation may be used:

$$\delta_{X-A} = \left[\left(\frac{\delta_{B-A}}{10^3} + 1 \right) \left(\frac{\delta_{X-B}}{10^3} + 1 \right) \right] \cdot 10^3 \tag{27}$$

where X represents the sample, A and B different standards.

Table 4. Absolute isotope ratios of international standards. (After Hayes 1983)

Standard	Ratio	Accepted value ($\times 10^6$) (with 95% confidence interval)	Source
SMOW	D/H	155.76± 0.10	Hagemann et al. (1970)
	$^{18}O/^{16}O$	2005.20± 0.43	Baertschi (1976)
	$^{17}O/^{16}O$	373±15	Nier (1950), corrected by Hayes (1983)
PDB	$^{13}C/^{12}C$	11 237.2 ± 2.9	Craig (1957)
	$^{18}O/^{16}O$	2067.1 ± 2.1	
	$^{17}O/^{16}O$	379±15	
Air nitrogen	$^{15}N/^{14}N$	3676.5 ± 8.1	Junk and Svec (1958)
Canyon Diablo Troilite (CDT)	$^{34}S/^{32}S$	45 004.5 ± 9.3	Jensen and Nakai (1962)

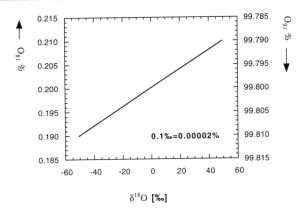

Fig. 10. Relationship between ^{18}O (^{16}O) content in percent and $\delta^{18}O$-value in per mill

$0.1‰=0.00002\%$

$\delta^{18}O$ **[‰]**

For different elements a convenient "working standard" is used in each laboratory. However, all values measured relative to the "working standard" are reported in the literature relative to a universal standard. As an example of the relationship between the content of an isotope in percent and the δ-value in per mill, Fig. 10 demonstrates that large changes in the δ-value only involve very small changes in the heavy isotope content (in this case the ^{18}O content). Unfortunately, there has not always been agreement among researchers in this field as to what standard should be designated as the universal standard. Such a standard should fulfill the following requirements:

1. Be used worldwide as the zero point.
2. Be homogeneous in composition.
3. Be available in relatively large amounts.
4. Be easy to handle for chemical preparation and isotopic measurement.
5. Have an isotope ratio near the middle of the natural variation range.

Among the reference samples now used, relatively few meet all of these requirements. For instance, the situation for the SMOW standard is rather confusing. The SMOW standard was originally a hypothetical water sample with an isotopic composition very similar to average ocean water (Craig 1961b), but defined in terms of a water sample distributed by the National Bureau of Standards (NBS-1). Later, the International Atomic Energy Authority (IAEA) distributed a water sample named V-SMOW (Vienna-SMOW), which is near but not the same in isotope composition to the original SMOW standard. The worldwide standards now in general use are given in Table 5.

The problems related to standards have been discussed by an IAEA advisory group, which met for the third time in 1983. As a result of these meetings (Coplen et al. 1983; Gonfiantini 1984), several new standards were agreed. A further advancement comes from interlaboratory comparison of two standards having different isotopic composition, such as the exercise organized by Blattner and Hulston (1978) on two carbonates. An interlaboratory calibration of this kind can then be used for a normalization procedure, which corrects for all proportional errors due to mass spectrometry and to sample preparation. Ideally, the two standard samples should have isotope ratios as different as possible, but still within

Table 5. Worldwide standards in use for the isotopic composition of hydrogen, carbon, oxygen, sulfur, and nitrogen

Element	Standard	Standard
H	Standard Mean Ocean Water	SMOW
C	Belemnitella americana from the Cretaceous Peedee formation, South Carolina	PDB
O	Standard Mean Ocean Water	SMOW
S	Troilite (FeS) from the Canyon Diablo iron meteorite	CD
N	Air nitrogen	N_2 (atm.)

the range of natural variations. There are, however, some problems connected with data normalization, which are still under debate. For example, the CO_2 equilibration of waters and the acid extraction of CO_2 from carbonates are indirect analytical procedures, involving temperature-dependent fractionation factors (whose values are not beyond experimental uncertainties) with respect to the original samples and which might be reevaluated on the normalized scale.

1.6
General Remarks on Sample Handling

Isotopic differences between samples to be measured are often extremely small. Therefore, great care has to be taken to avoid any isotope fractionation during chemical or physical treatment of the sample.

To convert geological samples to a suitable form for analysis, many different chemical preparation techniques must be used. These diverse techniques all have one general feature in common: any preparation procedure providing a yield of less than 100% may produce a reaction product that is isotopically different from the original specimen because the different isotopic species have different reaction rates.

A quantitative yield of a pure gas is usually necessary for the mass spectrometric measurement in order to prevent not only isotope fractionation during sample preparation, but also interference in the mass spectrometer. Contamination with gases having the same molecular masses and having similar physical properties may be a serious problem. This is especially critical with CO_2 and N_2O, on the one hand (Craig and Keeling 1963), and N_2 and CO, on the other. When CO_2 is used, interference by hydrocarbons and a CS^+ ion may also pose a problem.

Table 6. Gases most commonly used in isotope ratio in mass spectrometry

Element	Gas
H	H_2
C	CO_2
N	N_2
O	CO_2 (O_2)
S	SO_2, SF_6
Si	SiF_4

Contamination may result from incomplete evacuation of the vacuum system and/or from degassing of the sample. How gases are transferred, distilled, or otherwise processed in vacuum lines is briefly discussed in the sections dealing with the different elements. All errors due to chemical preparation limit the overall precision of an isotope ratio measurement to 0.1–0.2‰, while modern mass spectrometer instrumentation enables a precision better than 0.02‰ for light elements other than hydrogen. Larger uncertainties are expected when elements present in a sample at very low concentration are extracted by chemical methods (e.g., carbon and sulfur from igneous rocks). Table 6 summarizes which gases are used for mass-spectrometric analysis of the various elements.

1.7
New Microanalytical Techniques

In recent years microanalytical techniques, which permit relatively precise isotopic determinations on a variety of samples that are orders of magnitude smaller than those used in conventional techniques, have become increasingly important. Different approaches have been used in this connection.

1.7.1
Laser Microprobe

A recent summary has been given by Kyser (1995). Laser-assisted extraction is based on the fact that the energy of the laser beam is absorbed efficiently by the sample to be analyzed. The absorption characteristics depend on the structure, composition, and crystallinity of the sample. High-energy, finely focused laser beams have been used for some years in Ar isotope analysis, and the first well-documented preparation techniques with CO_2 and Nd:YAG laser systems for stable isotope determinations have been described by Crowe et al. (1990), Kelley and Fallick (1990), and Sharp (1990). Their results show that submilligram quantities of mineral can be analyzed for oxygen, sulfur, and carbon by transferring laser-volatilized gases (CO_2, SO_2) into a standard gas-source mass-spectrometer. Corrections have to be made for fractionations due to volatilization, which are necessary, mainly because of edge effects around the focused laser beam, where the sample is being only partly reacted. Recently, Wiechert and Hoefs (1995) described an excimer laser-based preparation technique which does not cause any fractionation during laser ablation due to edge or other effects.

1.7.2
Gas Chromatography Combined with Gas Mass Spectrometry

This technique has been summarized by Brand (1996). It employs a capillary column gas chromatograph and a combustion interface to produce CO_2 or N_2 interfaced with a modified isotope ratio mass spectrometer. The gas chromatograph is equipped with an on-column injector for fluids or a split injector for gases to ensure quantitative control over injections. Compounds eluting from the gas chromatograph are converted on-line into CO_2 and/or N_2 in the combustion interface. A modified type of mass spectrometer ensures the measurement of transient signals with high precision and the handling of carrier gas loads. As carrier gas, helium is utilized to introduce the sample gas into the ion source. In contrast to the dual-inlet system of conventional mass spectrometers, it is not possible to take the conventional sample for standard comparison and to measure the ion currents for a fixed time period. Instead the reference gas has to be introduced as if it were a sample peak. The gas containing the heavier isotopes precedes its lighter counterpart by 10–100 ms. Therefore, each peak must be integrated over its entire width to obtain the true isotope ratio. Standardization can be accomplished through the use of an added internal standard whose isotopic composition has been determined using conventional techniques.

1.7.3
Secondary Ion Mass Spectrometry

In an ion microprobe, secondary ions are produced by ion bombardment and focused into a fine beam. SIMS uses a Cs^+ primary beam to ablate the surface of a sample near the ion source of a mass spectrometer (Valley and Graham 1993; Eiler et al. (1995). The main advantages of this technique are its high sensitivity and its small sample size. Disadvantages are that the sputtering process produces a large variety of molecular secondary ions along with atomic ions which interfere with the atomic ions of interest and that the ionization efficiencies of different elements vary by many orders of magnitude and strongly depend on the chemical composition of the sample. This "matrix" effect is one of the major problems of quantitative analysis. Two instruments [Cameca and SHRIMP (sensitive high mass resolution ion micro probe)] have technical features, such as high resolving power and energy filtering, which help to overcome the problems of the presence of molecular isobaric interferences and the matrix dependence of secondary ion yields. As a result of these corrections, the precision of the SIMS technique is near the 1‰ level. SIMS has now been applied for different elements and minerals and some examples are given in Table 7.

The importance of these different microanalytical techniques cannot be overstated. The documentation of heterogeneity or homogeneity of geological samples at different scales can resolve many key questions in earth sciences.

Table 7. Applications of SIMS to different minerals

Element	Mineral	Reference
D/H	Amphiboles	Deloule et al. (1991)
$^{11}B/^{10}B$	Tourmalines	Chaussidon and Albarede (1992)
$^{15}N/^{14}N$	Graphite	Amari et al. (1993)
$^{13}C/^{12}C$	Diamonds	Harte and Otter (1992)
$^{18}O/^{16}O$	Magnetites	Valley and Graham (1991)
$^{34}S/^{32}S$	Sulfides	Eldridge et al. (1988)

CHAPTER 2

Isotope Fractionation Mechanisms of Selected Elements

The foundations of stable isotope geochemistry were laid in 1947 by Urey's classic paper on the thermodynamic properties of isotopic substances and by Nier's development of the ratio mass spectrometer. Before discussing details of the naturally occurring variations in stable isotope ratios, it is useful to describe some generalities that are pertinent to the field of isotope geochemistry as a whole.

1. Detectable isotope fractionation occurs only when the relative mass differences between the isotopes of a specific element are large. Therefore measurable isotope fractionations should be detectable only for the light elements (in general up to a mass number of about 40, see Table 8).

2. All elements that form solid, liquid, and gaseous compounds stable over a wide temperature range are likely to have variations in isotopic composition. Generally, the heavy isotope is concentrated in the solid phase in which it is more tightly bound. Heavier isotopes tend to concentrate in molecules in which they are present in the highest oxidation state.

3. Mass balance effects can cause isotope fractionations by changing modal proportions of substances during a chemical reaction. They are especially important for elements in situations where various reduced and oxidized compounds may coexist. Conservation of mass can be described by:

$$\delta_{(system)} = \Sigma_{i=1} \, x_i \, \delta_i \qquad (28)$$

where x_i is the mole fraction of the element in question for each phase within the system.

Table 8. Isotope abundances, relative mass differences and ranges of natural fractionations of selected elements (the reported range is not always identical with the maximum reported variation range, but is an estimate considered reasonable by the author)

Element	Isotope abundance Low mass		High mass		Mass difference (relative)	Range of natural fractionation (‰)
Hydrogen	^1H	99.984	^2D	0.016	2.00	400
Lithium	^6Li	7.52	^7Li	92.48	1.17	60
Boron	^{10}B	18.98	^{11}B	81.02	1.10	90
Carbon	^{12}C	98.89	^{13}C	1.11	1.08	100
Nitrogen	^{14}N	99.64	^{15}N	0.36	1.07	50
Oxygen	^{16}O	99.76	^{18}O	0.02	1.13	100
Silicon	^{28}Si	92.27	^{30}Si	3.05	1.07	5
Sulfur	^{32}S	95.02	^{34}S	4.21	1.06	150
Chlorine	^{35}Cl	75.53	^{37}Cl	24.47	1.06	15

4. Isotopic variations in most biological systems can be best explained by assuming kinetic effects. During biological reactions (e.g., photosynthesis, bacterial processes), the lighter isotope is very often enriched in the reaction product relative to the starting substances.

As a prelude to discussing the isotope characteristics of individual elements, a summary of isotope abundances, relative mass differences, and ranges of natural fractionations is provided in Table 8.

2.1
Hydrogen

Until 1931 it was assumed that hydrogen consisted of only one isotope. Urey et al. (1932a,b) detected the presence of a second stable isotope, which was called deuterium. (In addition to these two stable isotopes there is a third naturally occurring but radioactive isotope, ^3H, tritium, with a half-life of approximately 12.5 years). Way et al. (1950) gave the following average abundances of the stable hydrogen isotopes:

^1H: 99.9844%

^2D: 0.0156%

The isotope geochemistry of hydrogen is particularly interesting, for two reasons:

1. Hydrogen is omnipresent in terrestrial environments, occurring in different oxidation states in the forms of H_2O, OH-, H_2, and CH_4, even at great depths within the Earth. Therefore, hydrogen is envisaged to play a major role, directly or indirectly, in a wide variety of naturally occurring geological processes.

2. Hydrogen has by far the largest relative mass difference between its two stable isotopes. This results in hydrogen exhibiting the largest variations in stable isotope ratios of all elements.

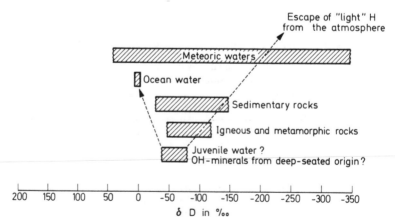

Fig. 11. D/H ratios of some geologically important materials (δD relative to SMOW)

The ranges of hydrogen isotope compositions of some geologically important reservoirs are given in Fig. 11. It is noteworthy that essentially all rocks on Earth have more or less the same hydrogen isotope composition, which is a characteristic feature of hydrogen, but not of the other elements. The reason for this overlap in isotope composition for rocks is likely due to the enormous amounts of water that have been cycled through the outer shell of the Earth.

2.1.1
Preparation Techniques and Mass Spectrometric Measurements

Determination of the D/H ratios is performed on H_2 gas. Water is converted to hydrogen by passage over hot uranium at about 750 °C, as described by Bigeleisen et al. (1952), Friedman (1953), and Godfrey (1962) or by passage over hot zinc at about 450 °C as described by Coleman et al. (1982). Most of the hydrogen generated from hydroxyl-bearing minerals is liberated in the form of water, but some is liberated as molecular hydrogen (Savin and Epstein 1970a). The resulting H_2 gas is generally converted to water by reaction with copper oxide. The water is then treated as described above.

A difficulty in measuring D/H isotope ratios is that, along with the H_2^+ and HD^+ formation in the ion source, H_3^+ is produced as a by-product of ion-molecule collisions. Therefore, a H_3^+ correction has to be made. The relevant procedures have been evaluated by Schoeller et al. (1983).

Analytical uncertainty for hydrogen isotope measurements is usually in the range ±0.5 to ±2‰, depending on different sample materials, preparation techniques, and laboratories.

2.1.2
Standards

There is a hierarchy of standards for hydrogen isotopes. The primary reference standard, the zero point of the δ-scale, is V-SMOW and SMOW, which are virtually identical in isotopic composition, the latter being a hypothetical water sample originally defined by Craig (1961b). The other standards, listed in Table 9, are used to verify the accuracy of sample preparation and mass spectrometry.

Table 9. Hydrogen isotope standards

Standards	Description	δ-value
V-SMOW	Vienna Standard Mean Ocean Water	0
GISP	Greenland Ice Sheet Precipitation	−189.9
V-SLAP	Vienna Standard Light Antarctic Precipitation	−428
NBS-30	Biotite	−65

2.1.3
Fractionation Mechanisms

The most effective processes in the generation of hydrogen isotope variations in
the terrestrial environment are phase transitions of water between vapor, liquid,

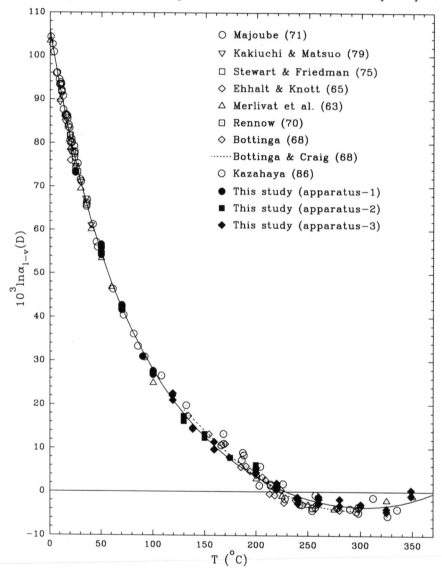

Fig. 12. Experimentally determined fractionation factors between H_2O_{liquid} and H_2O_{vapor} from 1 to
350 °C; for references see Horita and Wesolowski 1994. (After Horita and Wesolowski 1994; Reprinted from Geochimica et Cosmochimica Acta, Vol. 58 No. 16, Liquid-vapor fractionation of oxygen and hydrogen isotopes of water from the freezing to the critical temperature, pp 3425-3437,
© 1994, with kind permission from Elsevier Science Ltd, The Boulevard, Langford Lane, Kidlington OX5 1GB, UK)

and ice through evaporation/precipitation and boiling/condensation in the atmosphere, at the Earth's surface, and in the upper part of the crust. Differences in H-isotopic composition arise due to vapor pressure differences of water and, to a smaller degree, to differences in freezing points. Because the vapor pressure of HDO is slightly lower than that of H_2O, the concentration of D is lower in the vapor than in the liquid phase.

Horita and Wesolowski (1994) have summarized experimental results for the hydrogen isotope fractionation between liquid water and water vapor in the temperature range 0–350 °C (see Fig. 12). Hydrogen isotope fractionations decrease rapidly with increasing temperatures and become indistinguishable at 220–230 °C. Above the crossover temperature, water vapor is more enriched in deuterium than liquid water. Fractionations approach again to zero at the critical temperature of water (Fig. 12).

From experiments, Lehmann and Siegenthaler (1991) determined the equilibrium H-isotope fractionation between ice and water to be +21.2‰. Under natural conditions, however, ice will not necessarily be formed in isotopic equilibrium with the bulk water, depending mainly on the freezing rate.

In all processes concerning the evaporation and condensation of water, the hydrogen isotopes are fractionated in proportion to the oxygen isotopes, because a corresponding difference in vapor pressures exists between H_2O and HDO in one case and $H_2^{16}O$ and $H_2^{18}O$ in the other.

Therefore, the hydrogen and oxygen isotope distributions are correlated for meteoric waters. Craig (1961a) first defined the generalized relationship:

$$\delta D = 8\,\delta^{18}O + 10$$

which describes the interdependence of H- and O-isotope ratios in meteoric waters.

This relationship, shown in Fig. 13, has come to be described in the literature as the "meteoric water line (MWL)."

Neither the numerical coefficient 8 nor the constant 10, also called the deuterium excess d, are constant in nature. Both may vary depending on the conditions of

Fig. 13. Global relationship between monthly means of δD and $\delta^{18}O$ in precipitation, derived for all stations of the IAEA global network. *Line* indicates the global Meteoric Water Line (MWL). (After Rozanski 1993; Reprinted from Swart et al. (eds), Climate Change in Continental Isotopic Records, pp 1-36, © 1993, with kind permission from the American Geophysical Union, 2000 Florida Avenue, NW, Washington, DC 20009, U.S.A.)

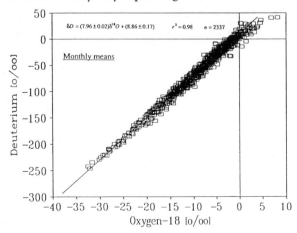

evaporation, vapor transport, and precipitation and, as a result, offer insight into climatic processes. A more detailed discussion of this feature is given in Sect. 3.6.

2.1.3.1
Equilibrium Exchange Reactions

D/H fractionations among gases are extraordinarily large, as calculated by Bottinga (1969a) and Richet et al. (1977) and plotted in Fig. 14. Even in magmatic systems, fractionation factors are sufficiently large to affect the δD-value of dissolved water in melts during degassing of H_2, H_2S, or CH_4. The oxidation of H_2 or CH_4 to H_2O and CO_2 is another process which due to the large fractionation factors may have an effect on the isotopic composition of water dissolved in melts.

The first set of experimentally determined hydrogen isotope fractionations between hydrous minerals and water was obtained by Suzuoki and Epstein (1976), who demonstrated the importance of the chemical composition of the octahedral sites in crystal lattices to the mineral H-isotope composition. Subsequent isotope exchange experiments by Graham et al. (1980, 1984) suggest that the chemical composition of sites other than the octahedral sites can also affect hydrogen isotope compositions. These authors postulate a qualitative relationship between hydrogen-bond distances and hydrogen isotope fractionations: the shorter the hydrogen bond, the more depleted the mineral is in deuterium.

Figure 15 summarizes H-isotope fractionation curves for several mineral–water systems, demonstrating that the forms of the curves can be extremely variable, even within one mineral group, and may also exhibit interesting features such as inflections, minima, and maxima.

When applying these experimental data to natural assemblages, isotope equilibrium is a necessary prerequisite, as noted previously. However, in the case of hydrogen, it is especially difficult to establish whether isotope equilibrium from high-temperature environments is preserved in a mineral during cooling, because the rate of H-isotope exchange is relatively rapid compared to oxygen iso-

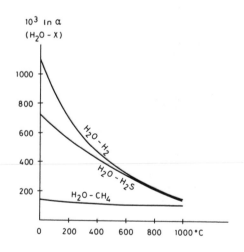

Fig. 14. D/H fractionations between H_2O and H_2, H_2O and H_2S, and H_2O and CH_4. (From calculated data of Richet et al. 1977)

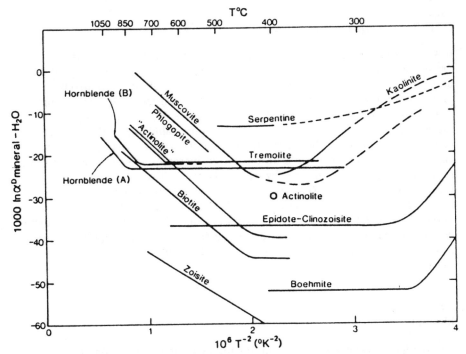

Fig. 15. Experimentally determined mineral-water hydrogen isotope fractionation curves. (Sheppard 1984)

topes. Graham (1981) demonstrated that mica closure temperatures for hydrogen are about 200 °C lower than those for oxygen. Rapid hydrogen transport may proceed by hydrolysis of Si–O and Al–O bonds, thus supporting the idea that water appears to be essential for isotope exchange. It was further demonstrated that hydrogen isotope exchange proceeds by a quite different mechanism in fluid-present versus fluid-absent systems. The presence of water greatly facilitates diffusion rates by at least two orders of magnitude.

2.1.3.2
Kinetic Isotope Effects

Hydrogen isotope fractionation occurs during photosynthesis such that deuterium is depleted in the organically bound hydrogen (White 1989). Owing to the complexity of the various hydrogen reactions occurring during cellular metabolism, quantitative modeling of individual steps involved in this fractionation has not yet been accomplished.

Appreciable hydrogen isotope fractionations seem probable in other biochemical processes as well, e.g., during bacterial production of molecular hydrogen and methane (Krichevsky et al. 1961). Cloud et al. (1958) observed that hydrogen gas given off by a bacterial culture was extremely depleted in deuterium.

2.1.3.3
Other Fractionation Effects

In salt solutions, isotopic fractionations can occur between the water in the "hydration sphere" and the free water (Truesdell 1974). The effects of dissolved salts on hydrogen isotope activity ratios in salt solutions can be qualitatively interpreted in terms of interactions between ions and water molecules, which appear to be primarily related to their charge and radius. Hydrogen isotope activity ratios of all salt solutions studied so far are appreciably higher than H-isotope composition ratios. As shown by Horita et al. (1993), the D/H ratio of water vapor in isotope equilibrium with a solution increases as salt is added to the solution. Magnitudes of the hydrogen isotope effects are in the order $CaCl_2 > MgCl_2 > MgSO_4 > KCl \sim NaCl > NaSO_4$ at the same molality.

The tendency for clays and shales to act as semipermeable membranes is well known. This effect is also known as "ultrafiltration." Coplen and Hanshaw (1973) postulated that hydrogen isotope fractionations may occur during ultrafiltration in such a way that the residual water is enriched in deuterium due to its preferential adsorption on the clay minerals.

2.2
Lithium

Lithium has two stable isotopes with the following abundances (Bainbridge and Nier 1950):

^{6}Li: 7.52%

^{7}Li: 92.48%

Lithium is one of the rare elements, where the lighter isotope is less abundant than the heavier one. This has led early authors to write δ-values as δ^6Li-values, which means that positive δ-values are isotopically "light" and, conversely, negative δ-values are isotopically "heavy," just the opposite to the general case. This is a cause for confusion, and it is reasonable to ask for a revision.

The relatively large mass difference between ^{6}Li and ^{7}Li of about 17% is a favorable condition for their fractionation in nature. Taylor and Urey (1938) found a change of 25% in the Li-isotope ratio when Li solutions percolate through a zeolite column. Thus, fractionation of Li isotopes might be expected in geochemical settings in which cation exchange processes are involved, perhaps during near-surface weathering processes. Because of the interesting nucleosynthesis of lithium, the determination of Li isotopes in extraterrestrial materials is potentially of great scientific interest.

As a result of serious instrumental fractionation effects during mass spectrometric analysis, the isotope geochemistry of lithium is still not very well known.

Some of these analytical difficulties have been overcome recently by analysis of lithium tetraborate (mass 56 and 57) instead of lithium directly (mass 6 and 7) (Chan 1987; Xiao and Beary 1989). Using this technique, Chan and Edmond (1988)

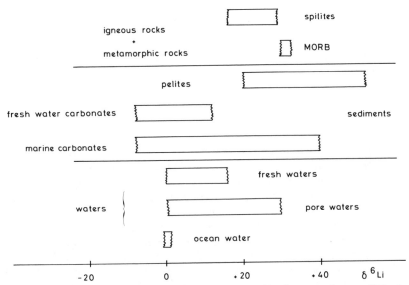

Fig. 16. Lithium isotope variations relative to ocean water (δ-values are given as δ⁶Li-values, which means that positive numbers are isotopically "light," negative numbers are isotopically "heavy." (Data from Sywall 1995)

reported a standard deviation of ±1.3‰, but to obtain a stable signal they had to use a relatively large sample size of 3 μg Li. Less precision (±3‰) but a higher sensitivity (10 ng Li) has been obtained by Sywall (1995) and Clausen (1995) by measuring masses 6 and 7 directly with a quadrupol mass spectrometer. From these measurements reported so far, it can be concluded that Li shows large isotope fractionations of up to 60‰ in the terrestrial environment (see Fig. 16).

Using ocean water as an external standard, mantle lithium has an isotope composition which is 30‰ lighter than ocean water (expressed as δ⁶Li-value +30‰). In this respect lithium isotope geochemistry shows strong similarities to that of boron. This isotopic difference between the mantle and oceanic reservoir is another powerful tracer that can be utilized to constrain water/rock interactions (Chan et al. 1992; You et al. 1995).

Very effective Li-isotope fractionation processes are operative in the sedimentary environment (Sywall 1995). Marine clays have a mean δ⁶Li-value of +28‰, whereas freshwater clays have lower δ-values. During weathering ⁷Li preferentially goes into solution, whereas ⁶Li is enriched in the weathering residue. The opposite effect is observed during clay mineral diagenesis (Sywall 1995).

Biogenic carbonates are somewhat variable in Li-isotope composition: while molluscs show a vital effect other organisms like corals do not show such a vital effect and obviously do not fractionate relative to ocean water.

Fossil marine carbonates exhibit a total Li-isotope variation from –10 to +50‰. By contrast, well-preserved carbonates have δ⁶Li-values which vary in a more restricted range from –7 to +32‰, so that it might be possible to deduce variations in the Li-isotope composition of fossil ocean water.

2.3
Boron

Boron has two stable isotopes with the following abundances (Bainbridge and Nier 1950):

^{10}B: 18.98%

^{11}B: 81.02%

The large relative mass difference between ^{10}B and ^{11}B and large chemical isotope effects between different species (Bigeleisen 1965) make boron a very promising element to study for isotope variations. However, boron is difficult to handle in mass-spectrometric measurements. The determination of gaseous boron compounds failed due to memory effects of BF_3.

In recent years solid source mass-spectrometry has provided an effective means for B-isotope analysis. Two different methods have been developed. The first was a positive thermal ionization technique using $Na_2BO_2^+$ ions initially developed by McMullen et al. (1961). Subsequently, Spivack and Edmond (1986) modified this technique by using $Cs_2BO_2^+$ ions. The substitution of ^{133}Cs for ^{23}Na increases the molecular mass and reduces the relative mass difference of its isotopic species, which limits the thermally induced mass-dependent isotopic fractionation. This latter method has a precision of about ±0.25‰, which is better by a factor of 10 than the $Na_2BO_2^+$ method. The other technique uses negative ionization producing BO_2^- ions (Vengosh et al. 1989). This technique has a lower precision of about ±2‰, but a considerably higher analytical sensitivity, thus allowing the measurement of boron samples in the nanogram range (Vengosh et al. 1989; Hemming and Hanson 1992). Yet another method has been used by Chaussidon and Albarede (1992), who performed boron isotope determinations with an ion microprobe having an analytical uncertainty of about ±2‰.

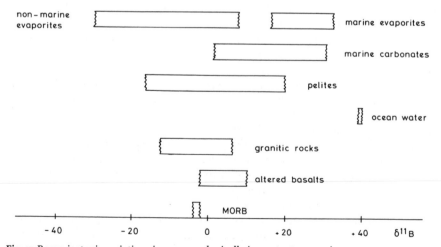

Fig. 17. Boron isotopic variations in some geologically important reservoirs

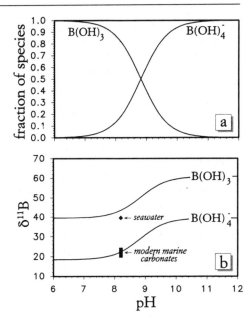

Fig. 18. a Distribution of aqueous boron species versus pH. **b** $\delta^{11}B$ of the two dominant species $[B(OH)_3, B(OH)_4^-]$ versus pH. (After Hemming and Hanson 1992; Reprinted from Geochimica et Cosmochimica Acta, Vol. 56 No. 1, Boron isotopic composition and concentration in modern marine carbonates, pp 537-543, © 1992, with kind permission from Elsevier Science Ltd, The Boulevard, Langford Lane, Kidlington OX5 1GB, UK)

$\delta^{11}B$-values are generally given relative to NBS boric acid SRM 951, which is prepared from a Searles Lake borax. This standard has a $^{11}B/^{10}B$ ratio of 4.04558 (Palmer and Slack 1989).

As analytical techniques have been consistently improved in recent years, the number of boron isotope studies has increased rapidly; nevertheless, boron isotope geochemistry is still in its infancy. Recent reviews have been given by Bassett (1990) and by Barth (1993). The total boron isotope variation documented to date is about 90‰. The lowest $\delta^{11}B$-values observed of around −30‰ are for certain tourmalines (Chaussidon and Albarede 1992) and some nonmarine evaporite sequences (Swihart et al. 1986), whereas the most enriched ^{11}B-reservoir is defined by brines from Australia and Israel (Dead Sea), which have $\delta^{11}B$-values up to 60‰ (Vengosh et al. 1991a,b). Isotope variations of boron in some geological reservoirs are shown in Fig. 17.

Boron is generally bound to oxygen or hydroxyl groups in either triangular (e.g., BO_3) or tetrahedral (e.g., $B(OH)_4^-$) coordination. The dominant isotope fractionation process occurs in aqueous systems via an equilibrium exchange process between boric acid $(B(OH)_3)$ and coexisting borate anion $(B(OH)_4^-)$. At low pH values trigonal $B(OH)_3$ predominates; at high pH values tetrahedral $B(OH)_4^-$ is the primary anion. Kakihana et al. (1977) have shown that the boric acid–borate fractionation is 23‰ at 25 °C, with the anion being depleted in ^{11}B. The pH dependence of the two boron species and their related isotope fractionation is shown in Fig. 18 (after Hemming and Hanson 1992).

The processes leading to the large ^{11}B-enrichment of seawater of about 40‰ relative to average continental material have been a matter of debate since first observed (Agyei and McMullen 1968; Schwarcz et al. 1969; Spivack and Edmond

1987; Palmer et al. 1987; Hemming and Hanson 1992). This enrichment has been attributed to isotopic fractionation associated with the absorption of boron onto clays, low-temperature alteration of oceanic crust, and carbonate minerals. The species preferentially absorbed on active mineral surfaces is the isotopically lighter $B(OH)_4^-$, thus shifting ocean water to higher $\delta^{11}B$-values. Because the relative abundances of boric acid and borate are a sensitive function of pH, B-isotope fractionations in water/rock reactions are pH dependent.

The isotope composition of boron in modern biogenic carbonate shells has been determined by Hemming and Hanson (1992) and Spivack et al. (1993). $\delta^{11}B$-values in carbonates are considerably lower than for seawater and suggest the preferential incorporation of the borate species into the carbonate lattice during precipitation. Thus, differences in ^{11}B contents of fossil carbonates might be due to differences in the pH value of the ocean in the past (Spivack et al. 1993; Sanyal et al. 1995). Ishikawa and Nakamura (1993) noted that ancient limestones are depleted in ^{11}B relative to recent carbonates and postulated that boron isotope fractionations must occur during diagenesis.

Tourmaline is the most abundant reservoir of boron in metamorphic and magmatic rocks. Swihart and Moore (1989) and Palmer and Slack (1989) analyzed tourmaline from various geological settings and observed a large range in $\delta^{11}B$-values from −22 to +22‰, which reflects the different origins of the boron (such as marine or nonmarine). Chaussidon and Albarede (1992) have studied tourmaline by ion microprobe and observed a systematic relationship between ^{11}B contents and chemical composition. Li-rich tourmalines are enriched in ^{11}B compared with Fe- and/or Mg-rich tourmalines. This relationship is interpreted to reflect variable contributions from marine and continental boron reservoirs.

Very interesting differences have been observed among basaltic rocks of different tectonic settings (Palmer 1991; Chaussidon and Jambon 1994; Chaussidon and Marty 1995). More positive $\delta^{11}B$-values of island arc volcanics relative MORB indicate the incorporation of altered ocean crust and subducted marine sediments into island arc magmas. $^{11}B/^{10}B$ ratios in volcanic glasses of oceanic island basalts are about 10‰ lighter than MORB (Chaussidon and Marty 1995). Since B isotope fractionations mainly occur at or near the surface of the Earth, $\delta^{11}B$ differences between MORB and ocean island basalts should somehow reflect surface processes (White, unpubl.).

2.4
Carbon

Carbon occurs in a wide variety of compounds on Earth, from highly reduced organic compounds in the biosphere to highly oxidized inorganic compounds such as CO_2 and carbonates. The broad spectrum of various oxidation states in a number of different geological settings is an ideal situation for naturally occurring fractionations.

Carbon has two stable isotopes (Nier 1950):

^{12}C: 98.89% (reference mass for atomic weight scale)

^{13}C: 1.11%

The naturally occurring variations in carbon isotope composition are greater than 100‰, neglecting extraterrestrial materials. Heavy carbonates with $\delta^{13}C$-values >+20‰ and light methane of <−80‰ have been reported in the literature.

2.4.1
Preparation Techniques

The gas used in all $^{13}C/^{12}C$ measurements is CO_2, for which the following preparation methods exist:

1. Carbonates are reacted with 100% phosphoric acid at temperatures between 20 and 75 °C (depending on the type of carbonate) to liberate CO_2 (see also Sect. 2.6).

2. Organic compounds are generally oxidized at high temperatures (850–1000 °C) in a stream of oxygen or by an oxidizing agent such as CuO. In the last few years, a new methodology to measure ^{13}C contents of individual compounds in complex organic mixtures has been developed. This so-called GC-C-MS technique, which employs a capillary column gas chromatograph, a combusion interface to produce CO_2, and a modified conventional gas mass-spectrometer, has the capability to measure individual carbon compounds in mixtures of subnanogram samples with a precision of better than ±0.5‰.

2.4.2
Standards

As the commonly used international reference standard PDB has been exhausted for several years, there is a need to introduce new standards. Several different standards are now in use; nevertheless the international standard the δ-values are referred to remains the PDB standard (Table 10).

2.4.3
Fractionation Mechanisms

The two main terrestrial carbon reservoirs, organic matter and sedimentary carbonates, have distinctly different isotopic characteristics because of the operation of two different reaction mechanisms:

1. Isotope equilibrium exchange reactions within the inorganic carbon system "atmospheric CO_2–dissolved bicarbonate–solid carbonate" lead to an enrichment of ^{13}C in carbonates.

2. Kinetic isotope effects during photosynthesis concentrate the light isotope ^{12}C in the synthesized organic material.

Table 10. $\delta^{13}C$-values of NBS-reference samples relative to PDB

NBS -18	Carbonatite	−5.00
NBS -19	Marble	+1.95
NBS -20	Limestone	−1.06
NBS -21	Graphite	−28.10

2.4.3.1
Inorganic Carbon System

The inorganic carbonate system comprises multiple chemical species linked by a series of equilibria:

$$CO_{2(aq)} + H_2O = H_2CO_3 \tag{1}$$

$$H_2CO_3 = H^+ + HCO_3^- \tag{2}$$

$$HCO_3^- = H^+ + CO_3^{2-} \tag{3}$$

The carbonate (CO_3^{2-}) ion can combine with divalent cations to form solid minerals, calcite and aragonite being the most common:

$$Ca^{2+} + CO_3^{2-} = CaCO_3 \tag{4}$$

An isotope effect is associated with each of these equilibria, the ^{13}C differences between the species depending only on temperature, although the relative abundances of the species are strongly dependent on pH. Figure 19 summarizes carbon isotope fractionations for carbonate–CO_2 and carbonate–HCO_3^- systems.

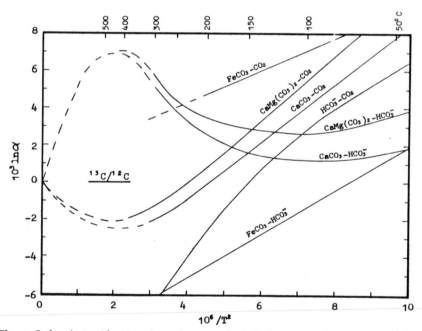

Fig. 19. Carbon isotope fractionations of some geologically important carbonate–CO_2 and carbonate–HCO_3^- systems as a function of temperature. (After Zheng and Hoefs 1993; Reprinted from Müller and Lüders (eds), Monograph Series on Mineral Deposits, Vol. 30, © 1993, with kind permission from Gebrüder Borntraeger Verlagsbuchhandlung, Johannesstraße 3A, 70176 Stuttgart, Germany)

At low temperatures, the largest fractionation occurs between dissolved CO_2 and bicarbonate. Several experimental studies have attempted to determine this equilibrium fractionation factor; of these the values determined by Mook et al. (1974) are considered the most reliable.

The generally accepted carbon isotope equilibrium values between calcium carbonate and dissolved bicarbonate are derived from inorganic precipitate data of Rubinson and Clayton (1969), Emrich et al. (1970), and Turner (1982). What is often not adequately recognized is the fact that systematic C-isotope differences exist between calcite and aragonite. Rubinson and Clayton (1969) found calcite and aragonite to be 0.9 and 2.7‰ enriched in ^{13}C relative to bicarbonate at 25 °C. Another complicating factor is the inability to demonstrate that shell carbonate – precipitated by marine organisms – is in isotopic equilibrium with the ambient dissolved bicarbonate. Such so-called "vital" effects can be as large as a few per mill (see discussion on p. 153).

Carbon isotope fractionations under equilibrium conditions are important not only under low-temperature conditions, but also at high temperatures within the system carbonate, CO_2, graphite, and CH_4. Of these, the calcite-graphite fractionation has become a useful geothermometer (i.e., Valley and O'Neil 1981; Scheele and Hoefs 1992; Kitchen and Valley 1995) (see discussion on p. 168).

2.4.3.2
Carbon Isotope Fractionations During Photosynthesis

Reviews by Deines (1980b) and O'Leary (1981) have summarized the biochemical background of carbon isotope fractionations during photosynthesis. The main isotope-discriminating steps during biological carbon fixation are (1) the uptake and intracellular diffusion of CO_2 and (2) the biosynthesis of cellular components. Such a two-step model was first proposed by Park and Epstein (1960):

$$CO_{2(external)} \overset{1}{\leftrightarrow} CO_{2(internal)} \overset{2}{\to} \text{organic molecule}$$

From this simplified scheme, it follows that the diffusional process is reversible, whereas the enzymatic carbon fixation is irreversible. Most plants use an enzyme called "ribulose bisphosphate carboxylase oxygenase (rubisco)," in which rubisco reacts with one molecule of CO_2 to produce three molecules of 3-phosphoglyceric acid. The carbon is subsequently reduced, carbohydrate formed, and the ribulose bisphosphate regenerated. Plants using this biochemical pathway are called C_3 plants and constitute about 90% of all plants today. C_3 plants generally have $\delta^{13}C$-values between –20 and –30‰. Another photosynthetic pathway is used by C_4 plants, which fix carbon by a process called phosphoenol pyruvate carboxylase (PEP). A much smaller C-isotope fractionation is associated with this biochemical pathway. A third group of plants is characterized by the "Crassulacean acid metabolism (CAM)," which can use both C_4 and C_3 metabolism, resulting in intermediate $\delta^{13}C$-values. (see also discussion in Sect. 3.10).

The two-step model of carbon fixation clearly suggests that isotope fractionation is dependent on the partial pressure of CO_2, i.e., pCO_2 of the system. With an unlimited amount of CO_2 available to a plant, the enzymatic fractionation will determine the ^{13}C content of photosynthetic carbon. Under these conditions, ^{13}C fractionations may vary from −17 to −40‰ (O'Leary 1981). When the concentration of CO_2 is the limiting factor, the diffusion of CO_2 into the plant is the slow step in the reaction and carbon isotope fractionations of the plant decrease.

C-isotope fractionation in aquatic plants is even more complex. As noted above, due to equilibrium isotope exchange HCO_3^- is enriched in ^{13}C relative to CO_2. Since HCO_3^- is much more abundant in seawater than dissolved CO_2, marine algae utilize ^{13}C-enriched HCO_3^- instead of CO_2, which explains why marine plants are generally enriched in ^{13}C relative to land plants.

Since the pioneering work of Park and Epstein (1960) and Abelson and Hoering (1961), it is well known that ^{13}C is not uniformly distributed among the total organic matter of plant material, but varies between carbohydrates, proteins, and lipids. The latter class of compounds is especially isotopically "light," i.e., they are depleted in ^{13}C relative to the other products of biosynthesis. Although the causes of these ^{13}C differences are not entirely clear, kinetic isotope effects seem to be more plausible (DeNiro and Epstein 1977; Monson and Hayes 1982) than thermodynamic equilibrium effects (Galimov 1985a,b). The latter author argued that ^{13}C concentrations at individual carbon positions within organic molecules are principally controlled by structural factors. Approximate calculations suggested that reduced C–H-bonded positions are systematically depleted in ^{13}C, while oxidized C–O-bonded positions are enriched in ^{13}C. Many of the observed relationships are qualitatively consistent with that concept; however, it is difficult to identify any general mechanism by which thermodynamic factors should be able to control chemical equilibrium within a complex organic structure. Experimental evidence presented by Monson and Hayes (1982) suggests that kinetic effects will be dominant in most biological systems.

Summarizing, the ^{13}C content of each biomolecule synthesized in nature depends on (1) the ^{13}C content of the carbon source, (2) isotope effects associated with the assimilation of carbon, (3) isotope effects associated with metabolism and biosynthesis, and (4) cellular carbon budgets.

2.4.4
Interactions Between the Carbonate-Carbon Reservoir and Organic Carbon Reservoir

Variations in ^{13}C content of some important carbon compounds are schematically demonstrated in Fig. 20. As shown in Fig. 20, the two most important carbon reservoirs on Earth, marine carbonates and the biogenic organic matter, are characterized by very different isotopic compositions: the carbonates being isotopically heavy with a mean $\delta^{13}C$-value around 0‰ and organic matter being isotopically light with a mean $\delta^{13}C$-value around −25‰. For these two sedimentary carbon reservoirs an isotope mass balance must exist such that:

$$\delta^{13}C_{input} = f_{org}\, \delta^{13}C_{org} + (1 - f_{org})\, \delta^{13}C_{carb} \tag{29}$$

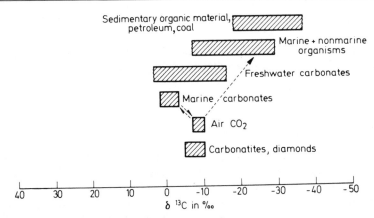

Fig. 20. δ¹³C-values of some important carbon reservoirs

If δ_{input}, δ_{org}, and δ_{carb} can be determined for a specific geological time, f_{org} can be calculated, where f_{org} is the fraction of organic carbon entering the sediments. It should be noted that f_{org} is defined in terms of the global mass balance and is independent of biological productivity, referring to the burial rather than the synthesis of organic material. That means that large f_{org} values might be a result of high productivity and average levels of preservation of organic material or of low levels of productivity and high levels of preservation.

The δ¹³C-value for the input carbon cannot be measured precisely but can be estimated with a high degree of certainty. As will be shown later, mantle carbon has an isotopic composition around –5‰ and estimates of the global average isotope composition for crustal carbon also fall in that range. Assigning a –5‰ value to δ¹³C input, f_{org} is calculated as 0.2 or in other words a ratio of $C_{org}/C_{carb} = 20/80$. As will be shown later (Sect. 3.8), f_{org} has obviously changed during specific periods of the Earth's history. Thus, knowledge of f_{org} is of great value in reconstructing the crustal redox budget.

2.5
Nitrogen

More than 99% of the known nitrogen on or near the Earth's surface is present as atmospheric N_2 or as dissolved N_2 in the ocean. Only a minor amount is combined with other elements, mainly C, O, and H. Nevertheless, this small part plays a decisive role in the biological world. Since nitrogen occurs in various oxidation states and in gaseous, dissolved, and solid forms (N_2, NO_3^-, NO_2^-, NH_3, NH_4^+), it is a highly suitable element for the search of natural variations in its isotopic composition. Schoenheimer and Rittenberg (1939) were the first to report nitrogen isotopic variations in biological materials. Today, the range of reported δ¹⁵N-values covers 100‰, from about –50‰ to +50‰; however, most δ-values fall within the much narrower spread from –10‰ to +20‰, as described in more recent reviews of the exogenic nitrogen cycle by Heaton (1986), Owens (1987), and Peterson and Fry (1987).

Nitrogen consists of two stable isotopes, ^{14}N and ^{15}N. Atmospheric nitrogen, determined by Nier (1950), has the following composition:

^{14}N: 99.64%

^{15}N: 0.36%.

N_2 is used for $^{15}N/^{14}N$ isotope ratio measurements, the standard being atmospheric N_2. Various preparation procedures have been described for the different nitrogen compounds (Bremner and Keeney 1966; Owens 1987; Velinsky et al. 1989; Kendall and Grim 1990; Scholten 1991 and others). In the early days of nitrogen isotope investigations, the extraction and combustion techniques potentially involved chemical treatments that could have introduced isotopic fractionations. In recent years, simplified techniques for combustion, such as commercial elemental analyzers, have come into routine use, so that a precision of 0.1‰–0.2‰ for $\delta^{15}N$ determinations can be achieved. Organic nitrogen-compounds are combusted to CO_2, H_2O, and N_2, the product gases are separated from each other cryogenically, and the purified N_2 is trapped on molecular sieves for mass-spectrometric analysis.

To understand the processes leading to the nitrogen isotope distribution in the geological environment, a short discussion of the biological nitrogen cycle is required. Atmospheric nitrogen can only be converted to organic nitrogen by certain bacteria and algae, which, in turn, are degraded to simple nitrogen compounds such as ammonium and nitrate. Thus, microorganisms are responsible for all major conversions in the biological nitrogen cycle, which generally is divided into fixation, nitrification, and denitrification.

Nitrogen fixation can be described by the reaction:

$$N_2 + 3H_2O \rightarrow 2NH_3 + 3/2\ O_2$$

and occurs in the roots of plants by many bacteria. The large amount of energy needed to break the molecular nitrogen bond makes nitrogen fixation a very inefficient process with little associated N-isotope fractionation.

The production of nitrate (nitrification) can be considered in terms of three different steps:

$$\text{organic N} \xrightarrow{1} NH_4 \xrightarrow{2} NO_2 \xrightarrow{3} NO_3$$

Step 1 involves very little fractionation, but step 2 (or steps 2 and 3) is accompanied by a large kinetic fractionation. The overall N-isotope fractionation for the whole process depends on which of the steps is rate limiting. If a relatively large amount of ammonium is available, step 2 or steps 2 and 3 become rate limiting, and the nitrate formed is depleted in ^{15}N by 20–35‰ (Mariotti et al. 1981). This is, however, not the rule. Most of the organic nitrogen in soils is slowly converted into ammonium and under these conditions step 1 becomes rate determining. Since

Table 11. Naturally observed isotope fractionation for nitrogen assimilation (after Fogel and Cifuentes 1995

N_2 fixation –3 to +1 ‰		
NH_4^+ Assimilation		
Cultures		
Millimolar concentrations	0 to –15 ‰	
Micromolar concentrations	–3 to –27 ‰	
Field observations		
Micromolar concentrations	–10 ‰	
NO_3^- Assimilation		
Cultures		
Millimolar concentrations	0 to –24 ‰	
Micromolar concentrations	–10 ‰	
Field observations		
Micromolar concentrations	–4 to –5 ‰	

this step is non-fractionating, the nitrate will have an isotope composition similar to that of its organic nitrogen source.

Denitrification (conversion of nitrate to N_2) takes place in poorly aerated soil and in stratified anaerobic water bodies. Denitrification supposedly balances the natural fixation of nitrogen; if it did not occur, then atmospheric nitrogen would be exhausted in less than 100 million years. A model for denitrification may involve two consecutive steps: (1) uptake of substrate into the cell with little or no isotope fractionation and (2) reduction of the substrate with the breaking of N–O bonds. This process occurs in a series of consecutive reaction steps and is associated with a large N-isotope effect (Mariotti et al. 1982). Experimental investigations have demonstrated that fractionation factors may change from 10 to 30‰, with the largest values obtained under the lowest reduction rates. Generally, the same factors that influence isotope fractionation during bacterial sulfate reduction are also operative during bacterial denitrification. Table 11, which gives a summary of observed N-isotope fractionations, clearly indicates the dependence of fractionations on nitrogen concentrations.

Thus far, only kinetic isotope effects have been considered, but isotopic fractionations associated with equilibrium exchange reactions have been demonstrated for the common inorganic nitrogen compounds (Letolle 1980). Of special importance in this respect is the ammonia volatilization reaction:

$$NH_{3gas} \leftrightarrow NH_{4\ aq}^+$$

for which isotope fractionation factors of 1.025–1.035 have been determined (Kirshenbaum et al. 1947; Mariotti et al. 1981). Experimental data by Nitzsche and Stiehl (1984) indicate fractionation factors of 1.0143 at 250 °C and of 1.0126 at 350 °C.

Many studies have shown that nitrogen isotopes can be used in environmental studies. Fertilizer, animal wastes, or sewage are the main sources of nitrate pollution in the hydrosphere. Under favorable conditions, these N-bearing materials can be isotopically distinguished from each other (Heaton 1986). Source studies also have been undertaken to trace the contribution of terrestrial organic matter to ocean water and to sediments (i.e., Sweeney et al. 1978; Sweeney and Kaplan 1980). Such studies are based, however, on the assumption that ^{15}N contents remain unchanged in the water column. Recent investigations by Cifuentes et al.

(1989), Altabet et al. (1991), and Montoya et al. (1991) have demonstrated there may be rapid temporal (even on a time scale of days) and spatial changes in the nitrogen isotope composition of the water column due to biogeochemical processes. This complicates a clear distinction between terrestrial and marine organic matter, although marine organic matter has generally a higher $^{15}N/^{14}N$ ratio than terrestrial organic matter.

Much of the initial organic nitrogen reaching the sediment/water interface is lost during early diagenesis. Nevertheless the nitrogen isotope composition of sediments is primarily determined by the source organic matter. In marine Cretaceous sediments, for instance, Rau et al. (1987) observed rather low ^{15}N contents which they interpreted to indicate an atypical marine biochemistry with reduced ocean circulation.

With further diagenesis of organic matter, nitrogen may also occur as ammonium incorporated in the lattice of clay minerals, where it replaces potassium. This nitrogen fixed in the crystal lattice of clay minerals and micas is mainly derived from decomposing organic matter and thus has a very similar isotopic composition to the organic matter (Scholten 1991; Williams et al. 1995).

During metamorphism of sediments, there is a significant loss of ammonium related to devolatilization, which is associated with a significant nitrogen fractionation, leaving behind ^{15}N-enriched residues (Haendel et al. 1986; Bebout and Fogel 1992; Boyd et al. 1993). Thus high-grade metamorphic rocks and granites are relatively enriched in ^{15}N and typically have $\delta^{15}N$-values between 8‰ and 10‰.

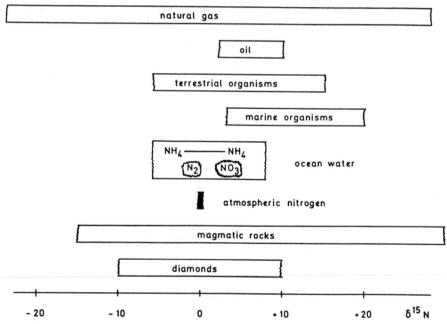

Fig. 21. $\delta^{15}N$-values relative to atmospheric nitrogen of some geologically important nitrogen compounds

The application of nitrogen isotopes in high-temperature studies (i.e., for elucidating processes such as fluid-rock exchange and devolatilization reactions) may have thus great potential but needs more detailed studies.

Figure 21 gives an overview of the nitrogen isotope variations in some important reservoirs.

2.6
Oxygen

Oxygen is the most abundant element on Earth. It occurs in gaseous, liquid, and solid compounds, most of which are thermally stable over large temperature ranges. These facts make oxygen one of the most interesting elements in isotope geochemistry.

Oxygen has three stable isotopes with the following abundances (Garlick 1969):

^{16}O: 99.763%

^{17}O: 0.0375%

^{18}O: 0.1995%

Because of the higher abundance and the greater mass difference, the $^{18}O/^{16}O$ ratio is normally determined, which may vary in natural samples by about 10% or in absolute numbers from about 1: 475 to 1: 525.

2.6.1
Preparation Techniques

In almost all laboratories, CO_2 is the gas used in the mass-spectrometric analysis. A wide variety of methods are used to liberate the oxygen from the various oxygen-containing compounds. Oxygen in silicates and oxides is usually converted to CO_2 through fluorination with F_2, BrF_5, or ClF_3 in nickel tubes at 500–650 °C (Taylor and Epstein 1962a; Clayton and Mayeda 1963; Borthwick and Harmon 1982). Decomposition by carbon reduction at 1000–2000 °C may be suitable for quartz and iron oxides but not for all silicates (Clayton and Epstein 1958). The oxygen is converted to CO_2 over heated graphite. Care must be taken to ensure quantitative oxygen yields, which can be a problem in the case of highly refractive minerals such as olivine and garnet. Low yields may result in anomalous $^{18}O/^{16}O$ ratios; high yields are often due to excess moisture in the vacuum extraction line.

Conventional fluorination is usually done on 10–20 mg of whole-rock powder or minerals separated from much larger samples. The inability to analyze small spots in situ means that natural heterogeneity cannot be detected. The recent development of laser microprobes, first described by Sharp (1990), has been a great advance in this respect. The laser evaporation has both the resolution and precision to investigate isotopic zoning within single mineral grains and mineral inter- and overgrowths.

Table 12. Isotopic fractionations for various carbonates
occuring during CO_2 liberation with phosphoric acid at
25 °C (mainly from Rosen-
baum and Sheppard 1986)

Carbonate	α	$10^3 \ln \alpha$
Calcite	1.01025	10.20
Aragonite	1.01034	10.29
Dolomite	1.01178	11.71
Siderite	1.01163	11.56

Carbonates are reacted with 100% phosphoric acid at various temperatures between 25 and 150 °C (McCrea 1950; Rosenbaum and Sheppard 1986; Swart et al. 1991). The following reaction equation:

$$3CaCO_3 + 2H_3PO_4 \rightarrow 3CO_2 + 3H_2O + Ca_3(PO_4)_2$$

shows that only two-thirds of the carbonate oxygen present in the product CO_2 is liberated. Since there are characteristic differences in the isotopic fractionation factors associated with the phosphoric acid liberation of CO_2 from various carbonates, this fact has to be considered when the isotopic composition of different carbonates is compared (see Table 12).

Experimental details of the phosphoric acid method vary significantly among different laboratories. The two most commonly varieties are the "sealed vessel" and the "acid bath" methods; in the latter method the CO_2 generated is continuously removed, while in the former it is not. Swart et al. (1991) demonstrated that the two methods exhibit a systematic ^{18}O difference between 0.2‰ and 0.4‰ over the temperature range 25–90 °C. Of these the "acid-bath" method probably provides the more accurate results. Wachter and Hayes (1985) demonstrated that careful attention must be given to the phosphoric acid. In their experiments best results were obtained by using a 105% phosphoric acid and a reaction temperature of 75 °C. This high reaction temperature should not be used when attempting to discriminate between mineralogically distinct carbonates by means of differential carbonate reaction rates.

Phosphates are first dissolved, then precipitated as ammonium phospho-molybdate and Mg phosphate to ensure purity and finally precipitated as Bi phosphate (Tudge 1960). The final Bi phosphate is then fluorinated. The procedure recently developed by Crowson et al. (1991) uses Ag_3PO_4 instead of Bi phosphate, which is less time consuming. O'Neil et al. (1994) developed a simpler method which eliminates purification steps and does not require the use of a fluorination system.

Sulfates are precipitated as $BaSO_4$, and then reduced with carbon at 1000 °C to produce CO_2 and CO. The CO is converted to CO_2 by electrical discharge between platinum electrodes (Longinelli and Craig 1967).

The $^{18}O/^{16}O$ ratio of water is usually determined by equilibration of a small amount of CO_2 with a surplus of water and analyzing the CO_2 after equilibration. For this technique the exact value of the fractionation for the $CO_2 = H_2O$ equilibrium at a given temperature is of crucial importance. A number of authors have experimentally determined this fractionation at 25 °C with variable results. A value of 1.0412 was proposed at the 1985 IAEA Consultants Group Meeting to be the best estimate.

It is also possible to quantitatively convert all water oxygen directly to CO_2 by reaction with guanidine hydrochloride. This technique was described by Dugan et al. (1985) and has the advantage that it is not necessary to assume a value for the H_2O–CO_2 isotope fractionation to arrive at a $^{18}O/^{16}O$ ratio.

2.6.2
Standards

Two different δ-scales are in use: $\delta^{18}O_{(SMOW)}$ and $\delta^{18}O_{(PDB)}$, because of two different categories of users, who have been traditionally engaged in O-isotope studies. The PDB standard is used in low-temperature carbonate studies. It is a Cretaceous belemnite from the Pee Dee formation and was the laboratory working standard used at the University of Chicago in the early 1950s when the paleotemperature scale was developed. The original supply of this standard has long been exhausted; therefore secondary standards have been introduced (see Table 13), whose isotopic compositions are considered calibrated to PDB. All other oxygen isotope analyses (waters, silicates, phosphates, sulfates, high-temperature carbonates) are given relative to SMOW.

The conversion equations of $\delta^{18}O_{(PDB)}$ versus $\delta^{18}O_{(SMOW)}$ and vice versa (Coplen et al. 1983) are:

$$\delta^{18}O_{(SMOW)} = 1.03091 \; \delta^{18}O_{(PDB)} + 30.91$$

and

$$\delta^{18}O_{(PDB)} = 0.97002 \; \delta^{18}O_{(SMOW)} - 29.98$$

Table 13 gives the $\delta^{18}O$-values of commonly used oxygen isotope standards on both scales (parentheses denote calculated values).

2.6.3
Fractionation Mechanisms

Of the numerous possibilities to fractionate oxygen isotopes in nature, the following are of special significance.

Knowledge of the oxygen isotope fractionation between liquid water and water vapor is essential for the interpretation of the isotope composition of different water types. Fractionation factors experimentally determined in the temperature range from 0 to 350 °C have been summarized by Horita and Wesolowski (1994).

Table 13. $\delta^{18}O$-values of commonly used O-isotope standards

Standard	Material	PDB scale	SMOW scale
NBS -19	Marble	-2.20	(28.64)
NBS -20	Limestone	-4.14	(26.64)
NBS -18	Carbonatite	-23.00	(7.20)
NBS -28	Quartz	(-20.67)	9.60
NBS -30	Biotite	(-25.30)	5.10
GISP	Water	(-53.99)	-24.75
SLAP	Water	(-83.82)	-55.50

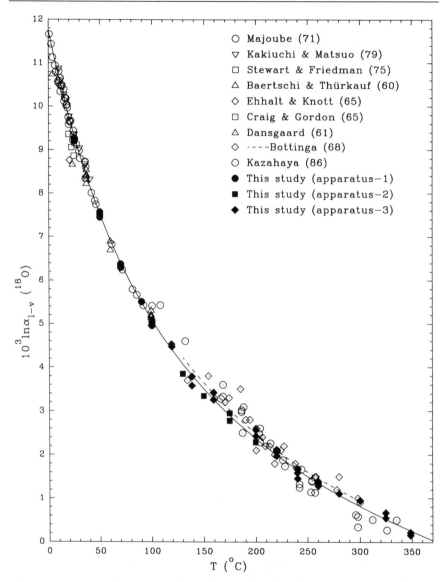

Fig. 22. Oxygen isotope fractionation factors between liquid water and water vapor in the temperature range 0–350 °C; for references see Horita and Wesolowski (1994). (After Horita and Wesolowski 1994; Reprinted from Geochimica et Cosmochimica Acta, Vol. 58 No. 16, Liquid-vapor fractionation of oxygen and hydrogen isotopes of water from the freezing to the critical temperature, pp 3425-3437, © 1994, with kind permission from Elsevier Science Ltd, The Boulevard, Langford Lane, Kidlington OX5 1GB, UK)

This is shown in Fig. 22. Upon the addition of salts to water, additional isotope effects may occur in water.

The presence of ionic salts in solution changes the local structure of water around dissolved ions. Taube (1954) first demonstrated that the $^{18}O/^{16}O$ ratio of CO_2 equilibrated with pure H_2O decreased upon the addition of $MgCl_2$, $AlCl_3$, and HCl, remained more or less unchanged for NaCl, and increased upon the addition of $CaCl_2$. The changes are roughly linear with the molality of the solute (see Fig. 23).

To explain this different isotopic behavior, Taube (1954) postulated different isotope effects between the isotopic properties of water in the hydration sphere of the cation and the remaining bulk water. The hydration sphere is highly ordered, whereas the outer layer is poorly ordered. The relative sizes of the two layers are dependent upon the magnitude of the electric field around the dissolved ions. The strength of the interaction between the dissolved ion and water molecules is also dependent upon the atomic mass of the atom to which the ion is bonded. O'-Neil and Truesdell (1991) have introduced the concept of "structure-making" and "structure-breaking" solutes: structure makers yield positive isotope fractionations whereas structure breakers produce negative isotope fractionations. Any solute that results in a positive isotope fractionation is one that causes the solution to be more structured as in the case of the ice structure, as compared to solutes that lead to less structured forms, in which cation–H_2O bonds are weaker than H_2O–H_2O bonds.

Of equal importance is the oxygen isotope fractionation in the CO_2–H_2O system. Most of the work has been concerned with the oxygen isotope partitioning

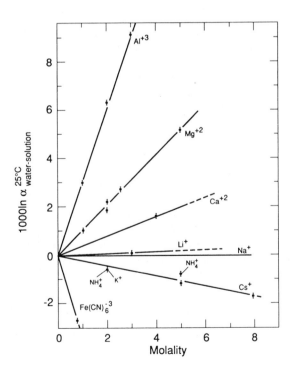

Fig. 23. Oxygen isotope fractionations between pure water and solutions of various ions. (After O'Neil and Truesdell 1991)

Table 14. Oxygen isotope fractionation factors relative water for the system $CO_2 - H_2O$ at 19 °C. (Usdowski and Hoefs 1993)

	$10^3 \ln \alpha$
H_2CO_3	38.7
HCO_3^-	34.5
CO_3^{2-}	18.2
$CO_{2(gas)}$	41.6
$CO_{2\,(aq.)}$	56.3

Table 15. Sequence of minerals in the order (bottom to top) of their increasing tendency to concentrate ^{18}O. The δ-values are hypothetical, but they are reasonably typical of low- to middle-grade metamorphism of pelitic schists. The minerals in parentheses are less well placed in the sequence than are the major minerals

Minerals	δ-value
1. Quartz	15.0
2. Dolomite	14.2
3. K-feldspar, albite	13.0
4. Calcite	12.8
5. Na-rich plagioclase	12.5
6. Ca-rich plagioclase	11.5
7. Muscovite, paragonite	11.3
8. Augite, orthopyroxene, diopside (kyanite glaucophane)	10.5
9. Hornblende (sphene, lawsonite)	10.0
10. Olivine, garnet (zircon, aopatite)	9.5
11. Biotite	8.5
12. Chlorite	8.0
13. Ilmenite	5.5
14. Magnetite, hematite	4.5

between gaseous CO_2 and water and gaseous CO_2 and carbonates. Usdowski et al. (1991) and Usdowski and Hoefs (1993) have calculated the fractionation factors for the individual carbonate species (Table 14).

The largest fractionation is caused by the oxygen isotope exchange between CO_2 and H_2O and the smallest fractionation occurs between dissolved and gaseous CO_2. The fractionation between the dissolved carbonate species and H_2O decreases in the order H_2CO_3, HCO_3^-, CO_3^{2-}.

The oxygen isotope composition of a rock depends on the ^{18}O contents of the constituent minerals and the mineral proportions. Gregory and Criss (1986) have reviewed the general principles of equilibrium isotope exchange among minerals in closed systems. They demonstrated that in a multimineralic rock, minor minerals will undergo a larger change in their δ-values than major minerals as a result of closed-system equilibration.

Taylor (1967) arranged coexisting minerals according to their relative tendencies to concentrate ^{18}O (Table 15).

This order of decreasing ^{18}O contents has been explained in terms of the bond type in the crystal structure. Semi-empirical bond-type calculations have been developed by Garlick (1966) and Savin and Lee (1988) by assuming that oxygen in a chemical bond has similar isotopic behavior regardless of the mineral in which the bond is located. This approach is useful for estimating fractionation factors. The accuracy of this approach is limited due to the assumption that the isotope fractionation depends only upon the atoms to which oxygen is bonded and not upon the structure of the mineral, which is not strictly true. By using an electrostatic approximation to bond strength and taking into account cation mass, Schütze (1980) developed an increment method for calculations of oxygen isotope fractionations in silicates, which has been modified and refined by Zheng (1991, 1993a,b).

On the basis of these systematic tendencies of ^{18}O enrichment found in nature, significant temperature information can be obtained up to temperatures of 1000 °C, and even higher, if calibration curves can be worked out for the various mineral pairs. The published literature contains many calibrations of oxygen isotope geothermometers, most having been determined by laboratory experiments, although some are based on theoretical calculations. As already discussed on p. 16, there are three sets of self-consistent fractionation factors for mineral pairs of geological interest: (1) semi-empirical calibrations, (2) theoretical calculations, and (3) experimental determinations.

Although much effort has been directed toward the experimental determination of oxygen isotope fractionation factors in mineral-water systems, the use of water as an oxygen isotope exchange medium has several disadvantages. Some minerals become unstable in contact with water at elevated temperatures and pressures, and lead to the occurrence of melting, breakdown, and hydration reactions. Incongruent solubility and ill-defined quench products may introduce additional uncertainties. Most of the disadvantages of water can be circumvented by using calcite as an exchange medium (Clayton et al. 1989; Chiba et al. 1989). Mineral-mineral fractionations – determined by these authors (Table 16) – give internally consistent geothermometric information that, generally, is in accord with independent estimates, such as the theoretical calibrations of Kieffer (1982).

Many isotopic fractionations between low-temperature minerals and water have been estimated by assuming that their temperature of formation and the isotopic composition of the water in which they formed (ocean water) are well known. This is sometimes the only approach available in cases in which the rates of isotope exchange reactions are slow and in which minerals cannot be synthesized in the laboratory at appropriate temperatures.

2.6.4
Fluid–Rock Interactions

Oxygen isotope ratio analysis provides a powerful tool for the study of water/rock interaction. The geochemical effect of such an interaction between water and rock or mineral is a shift of the oxygen isotope ratios of the rock and/or the water away from their initial values.

Table 16. Coefficients A for silicate – pair fractionations ($1000 \ln \alpha_{X-Y} = A/T^2 \ 10^6$). (After Chiba et al. 1989)

	Cc	Ab	An	Di	Fo	Mt
Qtz	0.38	0.94	1.99	2.75	3.67	6.29
Cc		0.56	1.61	2.37	3.29	5.91
Ab			1.05	1.81	2.73	5.35
An				0.76	1.68	4.30
Di					0.92	3.54
Fo						2.62

Detailed studies of the kinetics and mechanisms of oxygen isotope exchange between minerals and fluids show that there are three possible exchange mechanisms (Matthews et al. 1983b, c; Giletti 1985):

1. Solution-precipitation. During a solution-precipitation process, larger grains grow at the expense of smaller grains. Smaller grains dissolve and recrystallize on the surface of larger grains, which decreases the overall surface area and lowers the total free energy of the system. Isotopic exchange with the fluid occurs while material is in solution.

2. Chemical reaction. The chemical activity of one component of both fluid and solid is so different in the two phases that "a chemical reaction" occurs. The breakdown of a finite portion of the original crystal and the formation of new crystals is implied. The new crystals would form at or near isotopic equilibrium with the fluid.

3. Diffusion. During a diffusion process isotopic exchange takes place at the interface between the crystal and the fluid with little or no change in morphology of the reactant grains. The driving force is the random thermal motion of the atoms within a concentration or activity gradient.

Although diffusion is a relatively slow process by which isotope exchange takes place, it is likely to be the dominant isotope exchange mechanism in rocks and minerals. Although much faster, chemical reaction and recrystallization mechanisms typically will be of only minor importance.

The first attempts to quantify isotope exchange processes between water and rocks were made by Sheppard et al. (1969) and Taylor (1974). By using a simple closed-system material balance equation, these authors were able to calculate cumulative fluid/rock ratios:

$$W/R = \frac{\delta_{rock_f} - \delta_{rock_i}}{\delta_{H_2O_i} - (\delta_{rock_f} - \Delta)} \tag{30}$$

where $\Delta = \delta_{rock_f} - \delta_{H_2O_f}$.

Their equation requires adequate knowledge of both the initial (i) and final (f) isotopic states of the system and describes the interaction of one finite volume of rock with a fluid. The utility of such "zero-dimensional" or "one-box" equations has been recently questioned by Baumgartner and Rumble (1988), Blattner and Lassey (1989), Nabelek (1991), and others. Only under special conditions do one-box models yield information on the amount of fluid which actually flowed through the rocks. If the rock and the infiltrating fluid were not far out of isotopic equilibrium, then the calculated fluid/rock ratios rapidly approach infinity. Therefore, the equations are sensitive only to small fluid/rock ratios. Nevertheless, the equations can constrain fluid sources. More sophisticated models like the chromatographic or continuum mechanics models are physically more plausible, but give no information on the length of the interaction event.

Criss et al. (1987) and Gregory et al. (1989) developed a theoretical framework which describes the kinetics of oxygen isotope exchange between minerals and coexisting fluids. Figure 24 shows characteristic patterns of δ–δ plots in some hydrothermally altered granitic and gabbroic rocks. The $^{18}O/^{16}O$ arrays displayed on Fig. 24 cut across the 45° equilibrium lines at a steep angle as a result of the much faster oxygen isotope exchange of feldspar compared to that of quartz and pyroxene. If a low-^{18}O fluid such as meteoric or ocean water is involved in the exchange process, the slopes of the disequilibrium arrays can be regarded as "isochrons," in which the time increases as the slopes become less steep and approach the 45° equilibrium line. These "times" represent the duration of a particular hydrothermal event.

To conclude this section, Fig. 25 summarizes the naturally observed oxygen isotope variations in important geological reservoirs.

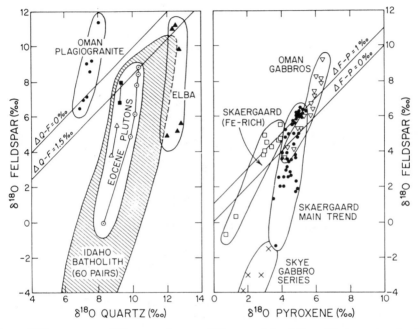

Fig. 24. $\delta^{18}O_{(feldspar)}$ versus $\delta^{18}O_{(quartz)}$ and versus $\delta^{18}O_{(pyroxene)}$ plots of disequilibrium mineral pair arrays in granitic and gabbroic rocks. These arrays indicate open-system conditions from circulation of hydrothermal fluids. (After Gregory et al. 1989)

Fig. 25. $\delta^{18}O$-values of important geological reservoirs

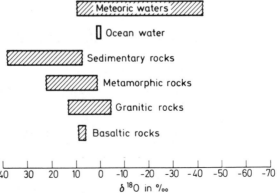

2.7
Silicon

Silicon has three stable isotopes with the following abundances (Bainbridge and Nier 1950):

^{28}Si: 92.27%

^{29}Si: 4.68%

^{30}Si: 3.05%

Because of its high abundance on Earth, silicon is in principle a very interesting element to study for isotope variations. However, because there is no redox geo-

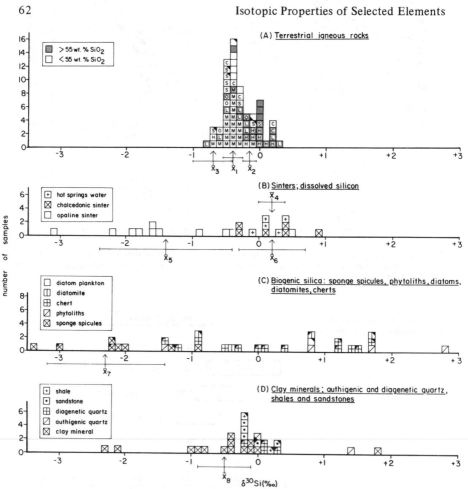

Fig. 26. Histogram of δ^{30}Si-values for various terrestrial samples (δ^{30}Si-values relative to CalTech Rose Quartz Standard). (Douthitt 1982)

chemistry for silicon and liquid or gaseous forms of silicon are of relatively little importance, only small isotope fractionations are to be expected in nature.

Douthitt (1982) has summarized the literature data and added new Si-isotope measurements on terrestrial materials. The total range of $\delta^{30}Si$ variation is 6.2‰ (see Fig. 26). In igneous rocks and minerals $\delta^{30}Si$-values show even smaller, but systematic variations, with ^{30}Si enrichment increasing with the silicon contents of igneous rocks and minerals. Relatively large fractionations occur in opaline sinters, biogenic opal, clay minerals, and authigenic quartz. A kinetic isotope fractionation of about 3.5‰ has been postulated by Douthitt (1982) to occur during the low-temperature precipitation of opal and possible, poorly ordered phyllosilicates. This fractionation coupled with a Rayleigh precipitation model is capable of explaining most non-magmatic $\delta^{30}Si$-variations. In a recent study, Jiang et al. (1994) postulated that the small differences observed in ^{30}Si content can be used in mineral exploration studies.

2.8
Sulfur

Sulfur has four stable isotopes with the following abundances (MacNamara and Thode 1950):

^{32}S: 95.02%

^{33}S: 0.75%

^{34}S: 4.21%

^{36}S: 0.02%

Sulfur is present in nearly all natural environments. It may be a major component in ore deposits, where sulfur is the dominant nonmetal, and as sulfates in evaporites. It occurs as a minor component in igneous and metamorphic rocks,

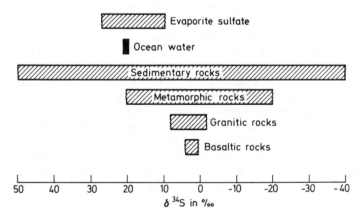

Fig. 27. $\delta^{34}S$-values of some geologically important materials ($\delta^{34}S$-values relative to Canyon Diablo troilite)

throughout the biosphere in organic substances, in marine sediments as both sulfide and sulfate, and in ocean water as sulfate. These occurrences cover the whole temperature range of geological interest. Thus, it is quite clear that sulfur is of special interest in stable isotope geochemistry.

Thode et al. (1949) and Trofimov (1949) were the first to observe wide variations in the abundances of sulfur isotopes. Variations on the order of 180‰ have been documented, with the "heaviest" sulfates having $\delta^{34}S$-values of greater than +120‰ (Hoefs, unpublished results), and the "lightest" sulfides having $\delta^{34}S$-values of around −65‰. Some of the naturally occurring S-isotope variations are summarized in Fig. 27. Previous reviews of the isotope geochemistry of sulfur have been published by Rye and Ohmoto (1974), Nielsen (1978, 1979), Ohmoto and Rye (1979), and Ohmoto (1986).

The reference standard commonly used is sulfur from troilite of the Canyon Diablo iron meteorite (CDT). As Beaudoin et al. (1994) have pointed out, CDT is not absolutely homogeneous and may display variations in ^{34}S up to 0.4‰.

2.8.1
Preparation Techniques

The chemical preparation of the various sulfur compounds for isotopic analysis has been discussed by Rafter (1957), Ricke (1964), Robinson and Kusakabe (1975) among others. The gas generally used in the mass-spectrometric measurement is SO_2, although Puchelt et al. (1971) and Rees (1978) describe a method using SF_6 which has some distinct advantages: it is without any mass-spectrometer memory effect and because fluorine is monoisotopic, no corrections of the raw data of measured isotope ratios are necessary.

Pure sulfides are converted to SO_2 by reaction with an oxidizing agent, such as CuO, Cu_2O, V_2O_5, or O_2. Combustion in vacuum with a solid oxidant minimizes the presence of contaminant gases, particularly CO_2, and purification of the SO_2 is often unnecessary. It is particularly important to minimize the production of sulfur trioxide since there is an isotope fractionation between SO_2 and SO_3. Special chemical treatment is necessary if pyrite is to be analyzed separately from other sulfides.

For the extraction of sulfates and total sulfur a suitable acid and reducing agent, such as tin(II)-phosphoric acid (the "Kiba" solution of Sasaki et al. 1979), is needed. The direct thermal reduction of sulfate to SO_2 has been described by Holt and Engelkeimer (1970) and Coleman and Moore (1978). Ueda and Sakai (1983) described a method in which sulfate and sulfide disseminated in rocks are converted to SO_2 and H_2S simultaneously, but analyzed separately.

In recent years, microanalytical techniques such as laser microprobe (Kelley and Fallick 1990; Crowe et al. 1990) and ion microprobe (Chaussidon et al. 1987, 1989; Eldridge et al. 1988, 1993) have become promising tools for determining sulfur isotope ratios. These techniques have several advantages over conventional techniques such as high spatial resolution and the capability for "in situ" spot analysis, thereby avoiding the difficulties in chemical preparation. However, sulfur isotopes are fractionated during ion or laser bombardment, but fractionation effects are mineral specific and reproducible.

2.8.2
Fractionation Mechanisms

Two types of fractionation mechanisms are responsible for the naturally occurring sulfur isotope variations:

1. A kinetic isotope effect during the bacterial reduction of sulfate, which produces the largest fractionations in the sulfur cycle.

2. Various chemical exchange reactions between both sulfate and sulfides and the different sulfides themselves.

2.8.2.1
Bacterial Reduction of Sulfate

The principal organisms which transform sulfate to hydrogen sulfide are anaerobic bacteria belonging to the genus *Desulphovibrio*, which gain their energy by coupling anaerobic oxidation of organic matter to the reduction of sulfate. Many of the environmental limitations of these bacteria were reviewed by Zobell (1958) and Chambers and Trudinger (1979). The rate of sulfate reduction is a function of a variety of parameters, the most important one being the reactivity of organic matter. The concentration of sulfate becomes important at rather low concentrations (less than 15% of the seawater value; Boudreau and Westrich 1984). The reaction chain during anaerobic sulfate reduction has been described in detail by Goldhaber and Kaplan (1974). In general, the rate-limiting step is the breaking of the first S–O bond, namely the reduction of sulfate to sulfite. Pure cultures of sulfate-reducing bacteria produce sulfide depleted in ^{34}S by 4‰–46‰ (Harrison and Thode 1957a,b; Kaplan et al. 1960; Kemp and Thode 1968; McCready et al. 1974; McCready 1975). In contrast, sulfides in sediments and euxinic waters are

Fig. 28. Rayleigh plot for sulfur isotope fractionations during the reduction of sulfate in a closed system. Assumed fractionation factor, 1.025; assumed starting composition of initial sulfate, +10‰

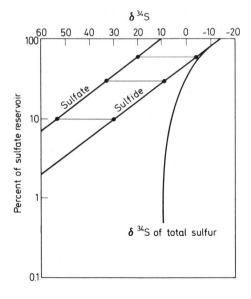

commonly depleted in ^{34}S by 45–70‰, far beyond the apparent capabilities of sulfate-reducing bacteria.

Most of the sulfide produced by sulfate reduction in sediments is reoxidized. Canfield and Thamdrup (1994) showed that through a repeated cycle of sulfide oxidation to elemental sulfur and subsequent disproportionation, bacteria can generate the large ^{34}S depletion of many marine sulfides. Thus the oxidative part of the sulfur cycle may create circumstances by which sulfides become more depleted in ^{34}S than would be possible with sulfate-reducing bacteria alone. Another factor which may influence the sulfur isotope composition is whether sulfate reduction takes place in an open or closed system.

An "open" system has an infinite reservoir of sulfate, in which continuous removal from the source produces no detectable loss of material. Typical examples are the Black Sea and local oceanic deeps. In such cases, H_2S is extremely depleted in ^{34}S while consumption and change in ^{34}S remain negligible. In a "closed" system, the preferential loss of the lighter isotope from the reservoir has a feedback on the isotopic composition of the unreacted source material. The changes in the ^{34}S content of residual sulfate and of the H_2S are modeled in Fig. 28, which shows that $\delta^{34}S$-values of the residual sulfate steadily increase with sulfate consumption (a linear relationship on the log-log plot). The curve for the derivative H_2S is parallel to the sulfate curve at a distance which depends on the magnitude of the fractionation factor. As shown in Fig. 28, H_2S may become isotopically heavier than the original sulfate when about two-thirds of the reservoir has been consumed. The $\delta^{34}S$-curve for "total" sulfide asymptotically approaches the initial value of the sulfate.

2.8.2.2
Thermochemical Reduction of Sulfate

In contrast to bacterial reduction, thermochemical sulfate reduction is an abiotic process, by which sulfate is reduced to sulfide under the influence of heat rather than bacteria (Trudinger et al. 1985; Krouse et al. 1988; Machel et al. 1995). The crucial question, which has been the subject of a controversial debate, is whether thermochemical sulfate reduction can proceed at temperatures as low as about 100 °C, just above the limit of microbiological reduction, given that laboratory experiments have failed to reduce sulfate abiotically (Trudinger et al. 1985). In recent years, there has been increasing evidence from natural occurrences that the reduction of aqueous sulfates by organic compounds could occur at temperatures as low as 100 °C, given enough time for the reduction to proceed (Krouse et al. 1988). Occurrences where thermochemical sulfate reduction has been suggested to be important are characterized by much smaller (or even no) isotope fractionations between sulfate and sulfide (or elemental sulfur).

2.8.2.3
Isotope Exchange Reactions

There have been a number of theoretical and experimental determinations of sulfur isotope fractionations between coexisting sulfide phases as a function of temperature. Theoretical studies of fractionations between sulfides have been undertaken by Sakai (1968) and Bachinski (1969), who reported the reduced partition function ratios and the bond strength of sulfide minerals and described the relationship of these parameters to isotope fractionation. In a manner similar to that for oxygen in silicates, there is a relative ordering of ^{34}S enrichment among coexisting sulfide minerals (Table 17). Considering the three most common sulfides (pyrite, sphalerite, and galena), under conditions of isotope equilibrium pyrite is always the most ^{34}S-enriched mineral and galena the most ^{34}S depleted; sphalerite displays an intermediate enrichment in ^{34}S.

Table 17. Equilibrium isotope fractionation factors of sulfides with respect to H_2S. The temperature dependence is given by A/T^2. (After Ohmoto and Rye 1979)

Mineral	Chemical composition	A
Pyrite	FeS_2	0.40
Sphalerite	ZnS	0.10
Pyrrhotite	FeS	0.10
Chalcopyrite	$CuFeS_2$	−0.05
Covelline	CuS	−0.40
Galena	PbS	−0.63
Chalcosite	Cu_2S	−0.75
Argentite	Ag_2S	−0.80

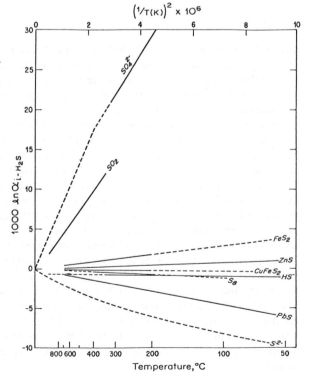

Fig. 29. Equilibrium fractionations among sulfur compounds relative to H_2S (*solid lines,* experimentally determined; *dashed lines,* extrapolated or theoretically calculated). (Ohmoto and Rye 1979)

The experimental determinations of sulfur isotope fractionations between various sulfides do not exhibit good agreement. The most suitable mineral pair for temperature determination is the sphalerite–galena pair. Rye (1974) has argued that the Czamanske and Rye (1974) fractionation curve gives the best agreement, with filling temperatures of fluid inclusions over the temperature range from 370 to 125 °C. By contrast, pyrite–galena pairs do not appear to be suitable for a temperature determination, because pyrite tends to precipitate over larger portions of ore deposition than galena, meaning that frequently these two minerals may not be contemporaneous. The equilibrium isotope fractionations for other sulfide pairs are generally so small that they are not useful as geothermometers. Ohmoto and Rye (1979) critically examined the available experimental data and presented a summary of what they believe to be the best S-isotope fractionation data. These S-isotope fractionations relative to H_2S are shown in Fig. 29.

2.9
Chlorine

Chlorine has two stable isotopes with the following abundances (Boyd et al. 1955):

^{35}Cl: 75.53%

^{37}Cl: 24.47%

Although there are no redox reactions, which induce a significant change in bonding energy, natural isotope variations in chlorine isotope ratios might be expected due to both the mass difference between ^{35}Cl and ^{37}Cl as well as to variations in coordination of chlorine among vapor, aqueous, and solid phases.

Measurements of Cl-isotope abundances have been made by different techniques. The first measurements by Hoering and Parker (1961) used gaseous chlorine in the form of HCl and the 81 samples measured exhibited no significant variations relative to the standard ocean chloride. In the early 1980s a new technique was developed by Kaufmann et al. (1984), which uses methylchloride (CH_3Cl). The chloride-containing sample is precipitated as AgCl, reacted with excess methyliodide, and separated by gas chromatography. The total analytical precision reported is near ±0.1‰ (Long et al. 1993; Eggenkamp 1994). The technique requires relatively large quantities of chlorine (>1 mg), which precludes the analysis of materials with low chlorine concentrations or which are limited in supply. A technique for the determination of Cl isotopes by negative thermal ionization has been described by Vengosh et al. (1989). This method involves no chemistry, but the precision is no better than ±2‰. Recently, Magenheim et al. (1994) described a method involving the thermal ionization of Cs_2Cl^+, which is both precise (±0.25‰) and sensitive, permitting the analysis of microgram quantities of chlorine. Chlorine is extracted from silicate samples via pyrohydrolysis, which is condensed in aqueous solution and then converted to CsCl by cation exchange.

Comprehensive studies of chlorine isotope variations using the methyliodide technique show relatively little isotope variation (Long et al. 1993; Eggenkamp 1994). The vast majority of both data sets have a comparable variation range from −1.4 to 1.5‰ relative to seawater chloride termed "SMOC" (standard mean ocean chloride). The largest isotopic differences have been found in slow-flowing groundwater, where Cl-isotope fractionation is attributed to a diffusion process (Kaufmann et al. 1984, 1988; Desaulniers et al. 1986). Chloride from fluid inclusions in hydrothermal minerals varies in the same range between −1.1‰ and 0.8‰ and indicates no significant differences between different types of ore deposits such as Mississippi Valley and Porphyry Copper type deposits (Eastoe et al. 1989; Eastoe and Gilbert 1992). Volpe and Spivack (1994) analyzed bulk marine aerosol samples collected from the atmosphere and found a 2‰ variation, which suggests that the atmosphere is isotopically inhomogeneous with respect to chlorine.

Much larger variations in the $^{37}Cl/^{35}Cl$ ratio have been observed by Magenheim et al. (1995) for the oceanic crust and its mantle source relative to seawater using the pyrohydrolysis technique. The $\delta^{37}Cl$-values for MORB glasses vary between 0.2 and 7.2‰. MORB samples which are unaffected by assimilation of hydrothermally altered oceanic crust have $\delta^{37}Cl$-values between 3 and 7‰. The data of Magenheim et al. (1995) seem to imply that chlorine is fractionated between the surface and the mantle and that a combination of degassing and recycling of oceanic crust during subduction is responsible for the observed differences in chlorine isotope composition.

Marine pore water associated with the Barbados and Nankai subduction zones shows, on the other hand, large negative $\delta^{37}Cl$-values down to −7.7‰ (Ransom et al. 1995). Chlorine, therefore, can no longer be regarded as being geochemically conservative in these systems and future studies may help in determining the extent to which chlorine may cycle between crust and mantle.

A summary of the so far observed natural chlorine isotope variations is presented in Fig. 30.

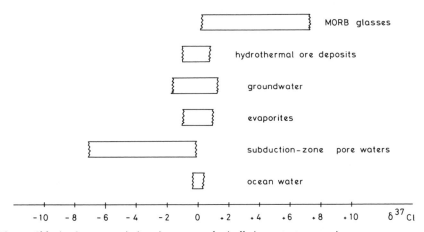

Fig. 30. Chlorine isotope variations in some geologically important reservoirs

CHAPTER 3

Variations of Stable Isotope Ratios in Nature

3.1
Extraterrestrial Materials

Information about nonterrestrial isotopic abundances can be obtained from studies of meteoritic, lunar, or planetary dust materials from both space probes and from ground-based astronomical observations. From these different sources of information it has been established that many elements have cosmic isotopic compositions which are very different from those observed in terrestrial material. One example of such large isotope variations is observed for primitive meteorites which contain minute interstellar dust grains of diamond, silicon carbide, and graphite that survived the formation of the solar system. Diamonds are too small to be analyzed as single grains, but silicon carbide and graphite have diameters in excess of 1 μm and can be analyzed isotopically by ion microprobe mass-spectrometry (Amari et al. 1993; Ott 1993). Silicon carbide and graphite grains exhibit variations in $^{12}C/^{13}C$ ratios that range up to more than three orders of magnitude greater than terrestrial C-isotope variations. Such large variations can be only explained in terms of several types of different stellar sources.

Isotope variations found in extraterrestrial materials have been classified according to different processes such as chemical mass fractionation, nuclear reactions, nucleosynthesis, and/or different sources such as interplanetary dust, solar materials, and comet material. For more detail the reader is referred to the reviews of Clayton (1993), Thiemens (1988), and others.

3.1.1
Meteorites

A major difference between extraterrestrial and terrestrial materials is the existence of primordial isotopic heterogeneities in the early solar system. These heterogeneities are not observed on the Earth or on the Moon, because they have become obliterated during high-temperature processes over geological time. In primitive meteorites, however, components that acquired their isotopic compositions through interaction with constituents of the solar nebula have remained unchanged since that time. The first observation which clearly demonstrated isotopic inhomogeneities in the early solar system was made by Clayton et al. (1973a). Previously, it had been thought that all physical and chemical processes must produce mass-dependent O-isotope fractionations yielding a straight line with a slope

Fig. 31. O-isotope composition of Ca–Al-rich inclusions (CAI) from carbonaceous chondrites and chondrules from carbonaceous, ordinary, and enstatite chondrites. (Clayton 1993; Reproduced, with permission , from the Annual Review of Earth and Planetary Sciences, Volume 21, © 1993, by Annual Reviews Inc.)

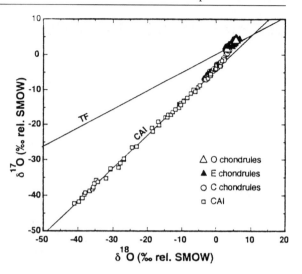

of 0.52 in a plot of ^{17}O versus ^{18}O. This line has been called the "terrestrial fractionation line." Figure 31 shows that O-isotope data from miscellaneous collections of terrestrial and lunar samples fall along the predicted mass-dependent fractionation line. However, selected anhydrous high-temperature minerals in carbonaceous chondrites, notably Allende, do not fall along the chemical fractionation trend, but instead define another trend with a slope of 1. Figure 31 shows oxygen isotope compositions for four groups of meteoritic samples. The first evidence for oxygen isotope anomalies was found in Ca–Al-rich refractory inclusions (CAI) in the Allende carbonaceous chondrite. These high-temperature inclusions are composed predominantly of melilite, pyroxene, and spinel. By analyzing various CAI-mineral fractions, Clayton et al. (1977) found a systematic relationship between O-isotopic composition and mineralogy. Spinel is the most ^{16}O-rich phase, with the lowest $\delta^{18}O$-value of about –40‰ and $\delta^{17}O$-value of about –42‰. Melilite, plagioclase, and Fe-bearing olivine are much less ^{16}O rich. The other group of samples in Fig. 31 represent chondrules from all classes of carbonaceous and ordinary chondrites. They cover a range of about 10‰ in $\delta^{18}O$ and $\delta^{17}O$. The pattern observed can be explained by mixtures of two components: one near the terrestrial composition, the other enriched in ^{16}O. The simplest model for the oxygen isotope composition of CAI and chondrules implies a common ^{16}O-rich dust component as the solid precursor and a ^{16}O-poor gas component with which the solid has exchanged during one or more heating events (Clayton 1993). Different nebular isotopic reservoirs must have existed, since there are distinct differences in bulk meteoritic O-isotope composition. On this basis, Clayton and Mayeda (1978) classified meteorites into seven categories in order of increasing ^{16}O contents: (1) L and LL chondrites, (2) H chondrites, (3) E chondrites and aubrites, (4) eucrites, howardites, and diogenites, (5) ureilites, (6) C2 carbonaceous chondrite hydrous matrix, and (7) C2, C3, and C4 carbonaceous chondrite anhydrous minerals.

Clayton and Mayeda (1978) have further tried to relate these seven meteorite groups to the various iron meteorite classes through O-isotope analysis of minor oxygen-bearing phases in iron meteorites in order to elucidate possible genetic links between stony and iron meteorites. The oxygen isotope observed distributions suggest that heterogeneities existed on scales ranging from submillimeters to hundreds of kilometers. The question remains as to where, when, and how the isotopic anomalies were originally produced (Thiemens 1988). If a chemical process is responsible for the bulk meteoritic oxygen isotope anomalies, then it is possible that only one reservoir could have produced the observed variations.

In addition to the oxygen isotope anomaly, the volatile elements H, C, N, and S also show extremely large variations in isotope composition in meteorites. In recent years most investigations have concentrated on the analyses of individual components, with more and more sophisticated analytical techniques. Of special interest is the analysis of meteoritic organic matter, because this may provide information about the origin of prebiotic organic matter in the early solar system. Two hypotheses have dominated the debate over formation mechanisms for the organic matter: (I) formation by a Fischer–Tropsch-type process promoted by catalytic mineral grains and (II) formation by Miller–Urey-type reactions in an atmosphere in contact with an aqueous phase. However, the isotopic variability exhibited by the volatile elements in different phases in carbonaceous chondrites is not readily compatible with an abiotic synthesis. Either complex variants of these reactions must be invoked, or totally different types of reactions need to be considered.

3.1.1.1
Hydrogen

The D/H ratio of the Sun is essentially zero: all the primordial deuterium originally present has been converted into ^{3}He during thermonuclear reactions. Therefore, analysis of primitive meteorites is the next best means of estimating the hydrogen isotope composition of the solar system.

In carbonaceous chondrites, hydrogen is bound in hydrated minerals and in organic matter. Considerable efforts have been undertaken to analyze D/H ratios of the different compounds (Robert et al. 1978, 1979a,b; Kolodny et al. 1980; Robert and Epstein 1982; Becker and Epstein 1982; Yang and Epstein 1983, 1984; Kerridge 1983; Kerridge et al. 1987; Halbout et al. 1990; Krishnamurthy et al. 1992). The δD-values in the phyllosilicates are relatively constant at around 110‰, whereas the organic matter shows exceptional deuterium enrichments, up to a δD-value of 10 000‰ in special deuterium carrier phases associated with the organic matter (Yang and Epstein 1983). The fact that the organic matter is D-rich confirms the long-held belief that it is indigenous to the meteorite. It also suggests that the organic substances are derived from interstellar precursors, an interpretation first proposed by Kolodny et al. (1980). Soluble organic molecules are also enriched in deuterium, but to a lesser extent, and their relationship to the bulk organic matter is unclear. Among the soluble organic compounds, amino acids exhibit the highest D/H ratios.

Neither the Fischer–Tropsch nor the Miller–Urey-type reactions are easy to reconcile with the high deuterium contents observed in extraterrestrial materials.

Therefore, it is necessary to consider an alternative mechanism for making D-rich organic compounds. A strong candidate for such an alternative mechanism is ion-molecule reactions that occur in interstellar space (Pillinger 1984).

3.1.1.2
Carbon

Besides the bulk carbon isotopic composition, the various carbon phases occurring in carbonaceous chondrites (kerogen, carbonates, graphite, diamond, silicon carbide) have been individually analyzed. The δ^{13}C-values of the total carbon fall into a narrow range, whereas δ^{13}C-values for different carbon compounds in single meteorites show extremely different ^{13}C contents. Figure 32 shows one such example, the Murray meteorite after Ming et al. (1989). Of special interest are the minute grains of silicon carbide and graphite in primitive carbonaceous chondrites, which obviously carry the chemical signature of the pre-solar environment (Ott 1993). The SiC grains, present at a level of a few parts per million, have a wide range in silicon and carbon isotope composition, with accompanying nitrogen also being isotopically highly variable. The ^{12}C/^{13}C ratio ranges from 2 to 2500, whereas it is 89 for the Earth. According to Ott (1993), the SiC grains can be regarded as "star dust," probably from carbon stars that existed long before our solar system. Amari et al. (1993) presented ion microprobe data of individual micrometer-sized graphite grains in the Murchison meteorite, also revealing large deviations from solar system values. These authors interpreted the isotope variability as indicating at least three different types of stellar sources. δ^{13}C-values reported for amino acids in the Murchison meteorite vary between +23 and +44‰ (Epstein et al. 1987). Engel et al. (1990) analyzed individual amino acids in the Murchison meteorite and also observed a strong ^{13}C enrichment. Of particular importance is the discovery of a distinct ^{13}C difference between D- and L-alanine, which suggests that optically active forms of material were present in the early solar system.

Fig. 32. Different carbon compounds in primitive meteorites. Species classified as interstellar on the basis of C-isotope composition are *shaded*. Only a minor, poorly known fraction of organic carbon is interstellar. (After Ming et al. 1989)

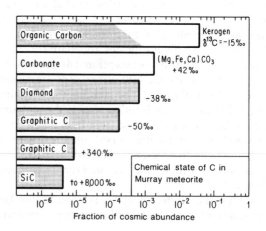

The ^{13}C enrichment in the carbonate has been attributed to kinetic isotope effects in Fischer–Tropsch-type reactions during the formation of organic compounds in these meteorites (Lancet and Anders 1970). However, Yuen et al. (1990) have questioned these findings from experiments in which the Fischer–Tropsch-type reactions were carried out in a dynamic rather a static system. Under such conditions, no large isotope fractionations were obtained between oxidized and reduced carbon.

3.1.1.3
Nitrogen

The nitrogen isotopes ^{14}N and ^{15}N are synthesized in two different astrophysical processes: ^{14}N during hydrostatic hydrogen burning and ^{15}N during explosive hydrogen and helium burning (Prombo and Clayton 1985). Thus, it can be expected that nitrogen should be isotopically heterogeneous in interstellar matter. What was considered by Kaplan (1975) to be a wide range of $\delta^{15}N$-values in meteorites has continuously expanded over the past 20 years (Kung and Clayton 1978; Robert and Epstein 1980, 1982; Lewis et al. 1983; Grady et al. 1985; Prombo and Clayton 1985). The lowest $\delta^{15}N$-value is $-326\permil$ in a fraction of an Allende acid residue; the highest $\delta^{15}N$-value is $973\permil$ in a whole-rock sample of the stony–iron meteorite of Bencubbin (Prombo and Clayton 1985). The $\delta^{15}N$-values for iron meteorites range from about -90 to about $+150\permil$ (Franchi et al. 1993). Most of the nitrogen is within the Fe,Ni metal phase, although occasionally high nitrogen concentrations can be found in nitrides. Small amounts of nitrogen may be also concentrated in phosphides and graphite. On the basis of nitrogen concentrations and isotope ratios, Franchi et al. (1993) divided iron meteorites into four groups. The large isotope variations may have been the result of parent body formation processes or may reflect variable conditions in the solar nebula during condensation and/or accretion.

3.1.1.4
Sulfur

There are many sulfur components in meteorites which may occur in all possible valence states (-2 to $+6$). Troilite is the most abundant sulfur compound of iron meteorites and has a relatively constant isotope composition (recall that troilite from the Canyon Diablo iron meteorite is the international sulfur standard, i.e., $\delta^{34}S$-value$=0\permil$). Carbonaceous chondrites contain sulfur of all valence states: sulfates, sulfides, elemental sulfur, and complex organic sulfur-containing molecules. Monster et al. (1965), Kaplan and Hulston (1966), and Gao and Thiemens (1993a,b) separated the various sulfur components and demonstrated that sulfides are characterized by the highest $\delta^{34}S$-values, whereas sulfates have the lowest $\delta^{34}S$-values, just the opposite to what is generally observed in terrestrial samples. This is strong evidence against any microbiological activity and instead a kinetic isotope fractionation in a sulfur-water reaction is favored (Monster et al.

1965). The largest internal isotope fractionation (7‰) is found in the Orgueil carbonaceous chondrite (Gao and Thiemens 1993a). Orgueil and Murchison show
internal isotopic variations between different specimens, which may indicate that
sulfur isotope heterogeneity existed in meteorite parent bodies. Since sulfur has
four stable isotopes, measurements of more than two isotopes may provide some
insights into nuclear processes and may help in identifying genetic relationships
between meteorites in a similar way to oxygen isotopes. However, early measurements by Hulston and Thode (1965) and Kaplan and Hulston (1966), and more recent ones by Gao and Thiemens (1991, 1993a,b), do not indicate any nuclear isotope anomaly and observed fractionations are mass dependent.

3.1.2
The Moon and the Neighboring Planets Mars and Venus

3.1.2.1
The Moon

Three different kinds of material are recognized on the lunar surface: (1) crystalline rocks of different chemical composition, (2) brecciated rocks from meteorite
impact, and (3) fines or soils with grain sizes down to 1 μm. The crystalline rocks
represent deep-seated lunar lithosphere, which is generally very poor in volatiles
carbon and nitrogen but rich in sulfur. The brecciated rocks represent an intermediate group, while the dust and fines have been heavily influenced through the
bombardment by the solar wind.

The oxygen isotope composition of the common igneous minerals is very constant with very little variation from one sampled locality to another (Onuma et al.
1970; Epstein and Taylor 1970, 1972; Clayton et al. 1973b). The $\delta^{18}O$-values for pyroxene are between 5.3 and 5.8‰, for olivines between 4.9‰ and 5.1‰, and for
plagioclases between 5.6 and 6.4‰. This small variation implies that the lunar in-
terior should have a $\delta^{18}O$-value of about 5.5‰, essentially identical to terrestrial
mantle rocks. The fractionations observed among coexisting minerals indicate
temperatures of crystallization of about 1000 °C or higher, similar to values observed in terrestrial basalts (Onuma et al. 1970). In comparison with other terrestrial rocks, the range of observed $\delta^{18}O$-values is very narrow. For instance, terrestrial plagioclases exhibit an O-isotope variation which is at least ten times greater than that for all lunar rocks (Taylor 1968). This difference may be attributed
to the much greater role of low-temperature processes in the evolution of the
Earths's crust and to the presence of water on the Earth.

The question of the presence of water on the Moon is very important in understanding the origin and conditions of its formation. Small amounts of water have
been found in lunar soils (Epstein and Taylor 1970, 1971, 1972; Friedman et al. 1970,
1974), but this water seems to be largely a result of terrestrial contamination.

The most notable feature of the sulfur isotope geochemistry of lunar rocks is
the uniformity of $\delta^{34}S$-values and their proximity to the Canyon Diablo standard.
The range of published $\delta^{34}S$-values is between –2 and +2.5‰; however, as noted
by Des Marais (1983), the actual range is likely to be considerably narrower than

4.5‰ due to systematic discrepancies either between laboratories or between analytical procedures. The very small variation in sulfur isotope composition sup-
ports the idea that the very low oxygen fugacities on the Moon prevent the formation of SO_2 or sulfate and thus eliminating exchange reactions between oxidized
and reduced sulfur species.

As further shown by Des Marais (1983), nitrogen and carbon abundances are
extremely low in lunar rocks. Des Marais presented compelling evidence that all
lunar rocks are contaminated by complex carbon compounds during sample
handling. This carbon, which is released at relatively low combustion temperatures, exhibits low $\delta^{13}C$-values, whereas the carbon liberated at higher temperatures has higher $^{13}C/^{12}C$ ratios. Another complication for the determination of
the indigenous isotope ratios of lunar carbon and nitrogen arises from spallation effects, which result from the interaction of cosmic ray particles with the lunar surface. These spallation effects lead to an increase in ^{13}C and ^{15}N, the extent
depending upon cosmic ray exposure ages of the rocks. Enrichments of the
heavy isotopes on the surfaces of the lunar fines are most probably due to the influence of the solar wind. Detailed interpretation of their isotopic variations is
difficult due to both the lack of knowledge of the isotopic composition of the solar wind and uncertainties of the mechanisms for trapping. Kerridge (1993)
demonstrated that nitrogen trapped in lunar surface rocks consists of at least
two components differing in release characteristics during experimental heating
and isotopic composition: the low-temperature component is consistent with
solar wind nitrogen, whereas the high-temperature component consists of solar
energetic particles.

3.1.2.2
Mars

Measurements from the Viking mission have shown that the Martian atmosphere
consists mainly of CO_2 with traces of N_2 and Ar. Oxygen and carbon isotopic composition are similar to those measured for the terrestrial atmosphere; however, a
strong enrichment in ^{15}N by about 75% relative to Earth has been found (Biemann
et al. 1976; Nier et al. 1976; Owen et al. 1977). This ^{15}N enrichment is attributed to
selective escape, implying a higher initial nitrogen abundance during the early
stages of Martian history.

Another way to study the Martian atmosphere was proposed by Wright et al.
(1990), who analyzed trapped gases in the SNC class of meteorites, commonly be-
lieved to come from Mars. These authors observed a 40‰ fractionation between
carbon associated with silicate minerals ($\delta^{13}C$: –25‰) and trapped CO_2 or carbonate minerals ($\delta^{13}C$: +15‰); they interpreted the relative ^{13}C enrichment in
surficial carbon as being due to the preferential removal of ^{12}C during an atmospheric loss process. Ion microprobe studies of amphibole, biotite, and apatite in
SNC meteorites by Watson et al. (1994) indicate high D/H ratios, with D enrichment extending up to five times the terrestrial value. This enormous degree of
deuterium enrichment can only be attributed to the escape of hydrogen to space,
as the lighter H isotope escapes so much more readily than D (Owen et al. 1988).

3.1.2.3
Venus

The mass spectrometer on the Pioneer mission in 1978 measured the atmospheric composition relative to CO_2, the dominant constituent. The $^{13}C/^{12}C$ and $^{18}O/^{16}O$ ratios were observed to be close to the Earth value, whereas the $^{15}N/^{14}N$ ratio was within 20% of that of the Earth (Hoffman et al. 1979). One of the major problems related to the origin and evolution of Venus is that of its "missing water." There is no liquid water on the surface of Venus today and the water vapor content in the atmosphere is probably not more than 220 ppm (Hoffman et al. 1979). This means that either Venus was formed of material very poor in water or whatever water was originally present has disappeared, possibly by the escape of hydrogen into space. And indeed Donahue et al. (1982) measured a 100-fold enrichment of deuterium relative to the Earth, which is consistent with such an outgassing process. The magnitude of this process is, however, difficult to understand.

3.1.3
Comets

Until 1985, knowledge of comets was based on observations obtained with telescopes. In March 1986 spacecraft missions obtained in situ data from the Halley comet, which have considerably expanded our knowledge about comets (i.e., review by Wyckoff 1991).

Isotopic ratio measurements in the Halley comet from the coma gases sampled in situ (D/H and $^{18}O/^{16}O$) and the bulk gas observed in ground-based spectra ($^{13}C/^{12}C$) provide insights into the early history of the cometary gases. The oxygen isotope ratio appears to be terrestrial, while D/H ratios indicate significant fractionation relative to the solar ratio. The carbon isotope ratio indicates a 30% excess in ^{13}C relative to the terrestrial isotope abundance.

3.1.4
Interplanetary Dust Particles

Interplanetary dust particles contribute about 10^4 tons/year to the Earth. They can be collected high in the Earth's stratosphere by aircraft at about 20 km altitude. Although these particles are small (typically 10 μm in diameter), they have been investigated by ion microprobe measurements (McKeegan 1987; McKeegan et al. 1985).

D/H ratios of individual dust particles give δD-values ranging from −386 to +2705‰, which thus exceed by far those in terrestrial samples. The hydrogen isotopic composition is heterogeneous on a scale of a few microns, demonstrating that the dust is unequilibrated. A carbonaceous phase rather than water appears to be the carrier of the D enrichment. Similar D enrichments in primitive meteorites have been attributed to the incorporation of deuterium into macromolecular organic matter.

In contrast to D/H ratios, an anomalous carbon isotope composition has not been clearly demonstrated. However, three of the investigated particles exhibit evidence for isotope heterogeneity. With respect to oxygen, none of the particles show large ^{16}O excesses of the type found in refractory oxide and silicate phases from carbonaceous chondrites.

These results suggest that the interplanetary dust particles are among the most primitive samples available for laboratory studies. Isotopically anomalous material constitutes only a small fraction of the investigated particles. Thus, it appears that the isotopic composition of these anomalous particles is not different from those observed in minor components of primitive meteorites.

3.2
Isotopic Composition of the Upper Mantle

Considerable geochemical and isotopic evidence has accumulated, supporting the concept that many parts of the mantle have experienced a complex history of partial melting, intrusion, crystallization, recrystallization, deformation, and metasomatism. A result of this complex history is that the mantle is chemically and isotopically heterogeneous.

Heterogeneities in radiogenic isotopes are relatively easy to detect because the processes which produce basaltic melts and a refractory residue do not fractionate radiogenic isotopes, or if this does occur the effects can be corrected by measurement of the non-radiogenic isotopes. Heterogeneities in stable isotopes are more difficult to detect: stable isotope ratios are affected by the various partial melting-crystal fractionation processes because they are governed by the temperature-dependent fractionation factors between residual crystals and partial melt and between cumulate crystals and residual liquid, but the magnitude of such effects is small at high temperature. Unlike radiogenic isotopes, stable isotopes are also fractionated by low-temperature surface processes. Therefore, they offer a potentially important means by which recycled crustal material can be distinguished from intra-mantle fractionation processes.

O,H,C,S, and N isotope compositions of mantle-derived rocks are substantially more variable than expected from the small fractionations expected at high temperatures. The most plausible process that may result in variable isotope ratios in the mantle is the input of oceanic crust, via subduction, into some portions of the mantle. This process should both release fluids rich in ^{18}O and D and variable in ^{13}C and ^{34}S content into the overlying mantle wedge and return compositionally and isotopically altered portions of the oceanic lithosphere into the deep mantle. In this context, the process of mantle metasomatism is of special significance. Metasomatic fluids rich in Fe^{3+}, Ti, K, LREE, P, and other LIL elements tend to react with peridotite mantle and form secondary micas, amphiboles, and other accessory minerals. The origin of metasomatic fluids is likely to be either (1) exsolved fluids from an ascending magma or (2) fluids or melts derived from subducted, hydrothermally altered oceanic crust.

With respect to the volatile behavior during partial melting, it should be noted that volatiles will be enriched in the melt and depleted in the parent material.

During ascent of melts, volatiles will be degassed preferentially, and this degassing will be accompanied by isotopic fractionation (see discussion in Sect. 3.4).

Sources of information about the isotopic composition of the upper portion of the lithospheric mantle are the direct analysis of unaltered ultramafic xenoliths brought rapidly to the surface in explosive volcanic vents. Due to rapid transport, these peridotite nodules are in many cases chemically fresh and considered by most workers to be the best sample available from the mantle. The other primary source of information is basalts, which represent partial melts from the mantle. The problem with basalts is that they do not necessarily represent the mantle composition because partial melting may have caused an isotopic fractionation relative to the precursor material. Partial melting of Ca–Al-containing peridotites would result in extraction of various basaltic magmas as the Ca–Al-rich minerals were dissolved, leaving behind refractory residues dominated by olivine and orthopyroxene which may slightly differ in the isotopic composition from the original materials. Also, basaltic melts may interact with the crustal lithosphere through which the magmas pass on their way to the Earth's surface. The following section will focus on ultramafic xenoliths; the isotopic characteristics of basalts are discussed in Section 3.3.

3.2.1
Oxygen

The $\delta^{18}O$-value of the bulk Earth is constrained by the composition of lunar basalts and meteorites to be in the region of 6‰. Insight into the detailed oxygen isotope character of the subcontinental lithospheric mantle has mostly come from the analysis of peridotitic xenoliths entrained in alkali basalts and kimberlites. The first oxygen isotope studies of such ultramafic nodules by Kyser et al. (1981, 1982) created much debate (e.g., Gregory and Taylor 1986a,b; Kyser et al. 1986). The Kyser et al. data showed that clinopyroxene and orthopyroxene had similar and rather constant $\delta^{18}O$-values around 5.5‰, whereas olivine exhibited a much broader ^{18}O variation, with $\delta^{18}O$-values extending from 4.5‰ to 7.2‰. Oxygen isotope fractionations between clinopyroxene and olivine (Δ_{cpx-ol}) vary from −1.4 to +1.2‰, implying that these phases are not in isotopic equilibrium at mantle temperatures. Gregory and Taylor (1986), through analogy with quartz–feldspar fractionations observed for hydrothermal systems, suggested that the disequilibrium fractionations in the peridotite xenoliths analyzed by Kyser et al. (1981, 1982) arose through open-system exchange with fluids having variable oxygen isotope compositions and with olivine exchanging ^{18}O more rapidly than pyroxene. This explanation remains to be validated, because it requires substantial amounts of fluid to bring about the ^{18}O shift in olivine and because self-diffusion in olivine is much slower than in pyroxene.

Furthermore, it should be recognized that olivine is a very refractory mineral and, as a result, quantitative reaction yields are generally not achieved, when analyzed by the convential fluorination technique. By contrast, the situation is distinctly different for the new laser fluorination technique. Mattey et al. (1994) analyzed 76 samples of olivine in spinel-, garnet-, and diamond-facies peridotites and observed an almost invariant O-isotope composition around 5.2‰. Assuming modal proportions of olivine, orthopyroxene, and clinopyroxene of 50: 40: 10, the calculated bulk mantle $\delta^{18}O$-value would be 5.5‰. Such a mantle source could generate liquids depending on melting temperatures and degree of partial melting, with O-isotope ratios equivalent to those observed for MORB and many ocean island basalts. These results do not rule out the existence of localized areas of the subcontinental mantle with a broader range in $\delta^{18}O$-values, produced as a result of metasomatic overprinting, but they argue that large domains of the mantle are more or less homogeneous in terms of their O-isotope composition. This conclusion is supported by recent investigations of Wiechert et al. (1996), who used an excimer-laser-based preparation technique, described by Wiechert and Hoefs (1995). Mineral separates from spinel peridotite xenoliths of Mongolia yielded the following mean values:

olivine: 5.41 ± 0.18

orthopyroxene: 5.90 ± 0.13

clinopyroxene: 5.71 ± 0.18

spinel: 5.08 ± 0.27

Standard deviations of the average values for the mantle minerals are somewhat above the reproducibility of the method (±0.1‰), which might indicate small O-isotope heterogeneities. However, some of the variation may be a result of cooling events, because larger crystals exhibit some ^{18}O zoning, with the outer rims being heavier than the cores.

Eclogite xenoliths from diamondiferous kimberlites constitute an important suite of xenoliths because they are among the deepest samples of the continental lithospheric mantle. Eclogite xenoliths have the most diverse range in $\delta^{18}O$-values between 2.2 and 7.9‰ (McGregor and Manton 1986; Ongley et al. 1987). This large range of ^{18}O variation indicates that the oxygen isotope composition of the continental lithosphere varies substantially, at least in any region where eclogite survives, and is the most compelling evidence that some nodules represent metamorphic equivalents of hydrothermally altered oceanic crust.

3.2.2
Hydrogen

The concept of "juvenile water" has a long tradition which has influenced thinking in various fields of igneous petrology and ore genesis. Juvenile water is defined as water that originates from degassing of the mantle and that has never been in contact with the Earth's surface. The analysis of OH-bearing minerals such as micas and amphiboles of deep-seated origin has been considered to be a source of information about juvenile water (e.g., Sheppard and Epstein 1970). Because knowledge about fractionation factors is limited and temperatures of final

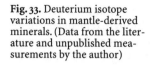

Fig. 33. Deuterium isotope variations in mantle-derived minerals. (Data from the literature and unpublished measurements by the author)

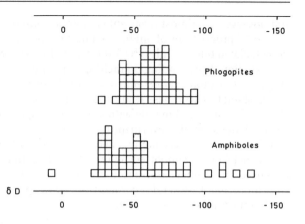

isotope equilibration between the minerals and water are not known, calculations of the H-isotope composition of the water in equilibrium with it are rather crude. Figure 33 gives δD-data on phlogopites and amphiboles, indicating that the hydrogen isotope composition of mantle water should lie in general between –80 and –50‰, the range first proposed by Sheppard and Epstein (1970) and subsequently supported by several other authors. Also shown in Fig. 33 are analyses for a considerable number of phlogopites and amphiboles which have δD-values higher than –50‰. Such elevated δD-values may indicate that water from subducted oceanic crust has played a role in the genesis of these minerals. Similar conclusions have been reached as a result of the analysis of water of submarine basalts from the Mariana arc (Poreda 1985) and from estimates of the original δD-values in boninites from Bonin Island (Dobson and O'Neil 1987).

An ion probe analytical procedure to measure in situ D/H ratios in amphiboles on the scale of a few tens of microns with a precision not better than ±10‰ has been described by Deloule et al. (1991). Samples analyzed included pargasites and kaersutite from ultramafic xenoliths and megacrysts of various localities. The observed δD-range from –125 to –12‰ is much larger than generally considered representative of mantle-derived samples. High δD-values from French Massif Central xenoliths may involve contamination by subduction of seawater-altered oceanic crust. Very low δD-values (as low as –125‰) for Hawaiian xenoliths were postulated to reflect the presence of a previously unidentified strongly D-depleted hydrogen component in the deep mantle.

3.2.3
Carbon

The presence of carbon in the upper mantle has been well documented through the following observations: CO_2 is a significant constituent in volcanic gases associated with basaltic eruptions with the dominant flux at mid-ocean ridges. The eruption of carbonatite and kimberlite rocks further testifies to the storage of CO_2 in the upper mantle. Additionally, the presence of diamond and graphite in kimberlites and eclogites reflects a wide range of mantle redox conditions, suggesting that carbon is related to a number of different processes in the mantle.

The isotopic composition of mantle carbon varies by more than 30‰ (see Fig. 34). To what extent this wide range is a result of mantle fractionation processes, the relict of accretional heterogeneities, or a product of the recycling of crustal carbon, is still unanswered (Mattey 1987).

In 1953, Craig noted that diamonds exhibited a range of δ^{13}C-values which clustered around –5‰. Subsequent investigations, which included carbonatites (e.g., Deines 1989) and kimberlites (e.g., Deines and Gold 1973), indicated similar δ^{13}C-values, which led to the concept that mantle carbon is relatively constant in C-isotopic composition, with δ^{13}C-values between –7‰ and –5‰. During the formation of a carbonatite magma, carbon is highly concentrated in the melt and is almost quantitatively extracted from its source reservoir. Since the carbon content of the mantle is low, the high carbon concentration of carbonatite melts requires extraction over volumes which may be some 10 000 times higher than the volume of a carbonatite magma (Deines 1989). Thus, the mean δ^{13}C-value of a carbonatite magma should represent the average carbon isotope composition of a relatively large volume within the mantle.

This situation is in total contrast with that for diamonds. As more diamond data became available (Deines et al. 1984, Galimov 1985 and others), the range of C-isotope variation broadened to more than 30‰. The present range in δ^{13}C is from –34 to +3‰ (Galimov 1991; Kirkley et al. 1991) and current debate centers around whether the more extreme values are characteristic of the mantle souce regions or whether they have resulted from isotope fractionation processes linked to diamond formation. The ^{13}C variability is not random but restricted to certain genetic classes. Common "peridotitic diamonds" have less variable carbon isotope compositions than "eclogitic diamonds," which span the entire

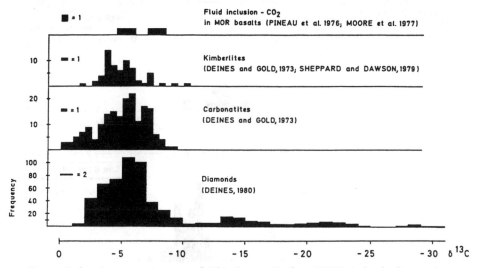

Fig. 34. Carbon isotope variations in fluid inclusion CO_2 from MORB, in kimberlites, carbonatites, and diamonds

range of $^{13}C/^{12}C$ variations observed to date. While some workers have argued that the variations are the result of high-temperature isotope fractionation processes within the mantle (Deines 1980a; Galimov 1991), others consider that peridotitic diamonds have formed from primitive carbon, whereas eclogitic diamonds have resulted from recycling of organic carbon (e.g., Kirkley et al. 1991).

3.2.4
Sulfur

Sulfur occurs in a variety of forms in the mantle, the major sulfur phase being monosulfide solid solution between Fe, Ni, and Cu. Recent ion microprobe analyses of sulfide inclusions from megacrysts and pyroxenite xenoliths from alkali basalts and kimberlites and in diamonds yielded $\delta^{34}S$-values from −11‰ to +14‰ (Chaussidon et al. 1987, 1989; Eldridge et al. 1991). Sulfur variations within diamonds exhibit the same characteristics as previously described for carbon, i.e., eclogitic diamonds are much more variable than peridotitic diamonds.

Interesting differences in sulfur isotope compositions are observed when comparing high-S peridotitic tectonites with low-S peridotite xenoliths (Fig. 35). Tectonites from the Pyrenees predominantly show negative $\delta^{34}S$-values around −5‰, whereas low-S xenoliths from Mongolia have largely positive $\delta^{34}S$-values up to +7‰. Ionov et al. (1992) determined sulfur contents and isotopic compositions in some 90 garnet and spinel lherzolites from six regions in southern Si-

Fig. 35. Sulfur isotope composition of high- and low-S peridotites (Pyrenaen tectonites, Chaussidon and Lorand 1990; Mongolian xenoliths, Ionov et al. 1992)

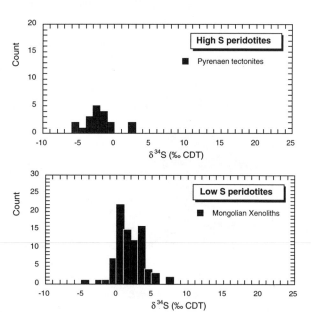

beria and Mongolia for which the range observed in $\delta^{34}S$ was from $-7‰$ to $+7‰$. Ionov et al. (1992) concluded that low sulfur concentrations (<50 ppm) and largely positive $\delta^{34}S$-values predominate in the lithospheric continental mantle worldwide. S-isotope compositions typical of MORB ($\delta^{34}S$: $0-1‰$) may be produced by melting of moderately depleted lherzolites. Primary melts with positive $\delta^{34}S$-values may be generated from mantle peridotites with larger degrees of depletion and/or from rocks metasomatized by subduction-related fluids.

3.2.5
Nitrogen

Because of the inert nature of nitrogen, it might be expected that the nitrogen isotopic composition of the mantle would be similar to that of the atmosphere. This is, however, not the case. Diamonds provide the most important source of information about mantle nitrogen, because nitrogen is their main trace component. $\delta^{15}N$-values vary widely from about -12 to $+12‰$ (Javoy et al. 1986; Boyd et al. 1992; Boyd and Pillinger 1994). ^{15}N differences of up to $10‰$ have been observed within individual diamonds, with cores being enriched in ^{15}N relative to coats (Boyd et al. 1992). A coupling of $\delta^{13}C$ with $\delta^{15}N$ values has been demonstrated by Boyd and Pillinger (1994): nitrogen in "high-$\delta^{13}C$" diamonds is generally depleted in ^{15}N, whereas nitrogen in "low-$\delta^{13}C$" diamonds is enriched in ^{15}N. This finding is consistent with the explanation of subduction of crustal material; within mantle isotope fractionations cannot be ruled out, however (Boyd and Pillinger 1994).

3.3
Magmatic Rocks

Magmatic rocks should exhibit relatively small differences in isotopic composition because of their high temperature of formation. However, primarily as a result of secondary alteration processes, the variation observed in isotopic composition of magmatic rocks can be much greater than expected from their temperature of formation.

Provided an igneous rock has not been affected by subsolidus isotope exchange or hydrothermal alteration, its isotope composition will be determined by:

1. The isotope composition of the source region in which the magma was generated
2. The temperature of magma generation and crystallization
3. The mineralogical composition of the rock
4. The evolutionary history of the magma including such additional processes as isotope exchange, assimilation of country rocks, and magma mixing

In the following sections, which concentrate on $^{18}O/^{16}O$ measurements, some of these points are discussed in more detail (see also Taylor 1968, 1986; Taylor and Sheppard 1986).

3.3.1
Fractional Crystallization

Because fractionation factors between liquid and solid are small at the relatively high temperatures of magmatic melts, fractional crystallization seems to play only a minor role in influencing the oxygen isotopic composition of magmatic rocks. Matsuhisa (1979) reported that within a lava sequence from Japan $\delta^{18}O$ increased by approximately 1‰ from basalt to dacite. Muehlenbachs and Byerly (1982) analyzed an extremely differentiated suite of volcanic rocks at the Galapagos spreading center and showed that 90% fractionation only enriched the residual melt by about 1.2‰. On Ascension Island, Sheppard and Harris (1985) observed a difference of nearly 1‰ in a volcanic suite ranging from basalt to obsidian. In summary, modeling for closed-system, crystal fractionation predicts an ^{18}O enrichment of about 0.4‰ per 10 wt% increase in SiO_2 content.

3.3.2
Differences Between Volcanic and Plutonic Rocks

Systematic differences in O-isotope composition are observed between fine-grained, rapidly quenched volcanic rocks and their coarse-grained plutonic equivalents (Taylor 1968; Anderson et al. 1971). Fractionations among minerals in plutonic mafic rocks are on average about twice as great as for the corresponding fractionations observed in extrusive mafic rocks. This difference may result from the retrograde exchange or post-crystallization exchange reactions of the plutonic rocks with a fluid phase. This interpretation is supported by the fact that basaltic and gabbroic rocks from the lunar surface yield the same "isotopic temperatures" corresponding to their initial temperatures of crystallization. Due to the absence of water on the moon, no retrograde exchange took place.

3.3.3
Low-Temperature Alteration Processes

Because of their high glass contents and very fine grain size, volcanic rocks are very susceptible to low-temperature processes such as hydration and weathering, which are characterized by large ^{18}O enrichment effects. In general, it is probable that Tertiary and older volcanic rocks will exhibit O-isotope compositions that have been modified to higher $^{18}O/^{16}O$ ratios from their primary state to some extent (Taylor 1968; Muehlenbachs and Clayton 1972; Cerling et al. 1985; Ferrara et al. 1986; Harmon et al. 1987). Although there is no way of exactly ascertaining the magnitude of these ^{18}O enrichments on a sample by sample basis, a crude estimate can be made by determining the water (and carbon dioxide) content and "correcting" to what are considered primary values of the suite of rocks to be analyzed (Taylor et al. 1984; Ferrara et al. 1985; Harmon et al. 1987). The primary water content of a magma is difficult to estimate, but it is generally accepted that primary basaltic magmas should not contain more than 1% water. Thus, any water content >1% should be of secondary origin and the $\delta^{18}O$-value for such samples should be corrected before such ^{18}O measurements can be used for petrogenetic interpretations.

3.3.4
Assimilation of Crustal Rocks

Because the various surface and crustal environments are characterized by different and distinctive isotope compositions, stable isotopes provide a powerful tool of discriminating between the relative role of mantle and crust in magma genesis. This is especially true when stable isotopes are considered together with radiogenic isotopes, because variations within these independent isotopic systems arise from unrelated geological causes. For instance, a mantle melt that has been affected by contamination processes within the upper crust will exhibit increases in $^{18}O/^{16}O$ and $^{87}Sr/^{86}Sr$ ratios that correlate with increase in SiO_2 and decrease in Sr content. In contrast, a mantle melt which evolves only through differentiation unaccompanied by interaction with crustal material will have an O-isotope composition which mainly reflects that of its source region, independent of variations in chemical composition. In this latter case, correlated stable and radiogenic isotope variations would be an indication of variable crustal contamination of the source region (i.e., crustal material which has been recycled into the mantle via subduction).

Modeling by Taylor (1980) and James (1981) has demonstrated that it is possible to distinguish between the effects of source contamination as well as crustal contamination. Magma mixing and source contamination are two-component mixing processes which obey two-component hyperbolic mixing relations, whereas crustal contamination is a three-component mixing process, involving the magma, the crustal contaminant, and the cumulates, that results in more complex mixing trajectories on an oxygen–radiogenic isotope plot.

3.3.5
Basaltic Rocks from Different Tectonic Settings

Harmon and Hoefs (1995) have assembled a database consisting of 2855 O-isotope analyses of Neogene volcanic rocks worldwide. The total oxygen isotope variation is almost 18‰, extending from $\delta^{18}O$-values of –0.2‰ to 17.4‰ and having a mean value of 6.8‰. To remove from this sample population lavas whose O-isotope compositions had been affected by secondary subsolidus alteration, Harmon and Hoefs (1995) have screened their database to include only (1) historic lavas younger than 1500 years B.P., (2) pristine submarine glasses, and (3) samples with measured water contents of less than 0.75% H_2O. This screening procedure reduced the average $\delta^{18}O$-value only from 6.8 to 6.2, but with the observed variation range for the 743 basalts still exhibiting a substantial ^{18}O-variation from 2.9 to 11.4‰.

To understand the implications of this large variation in basaltic rocks, it is illustrative to consider basalt $\delta^{18}O$ variations as a function of bulk compositions

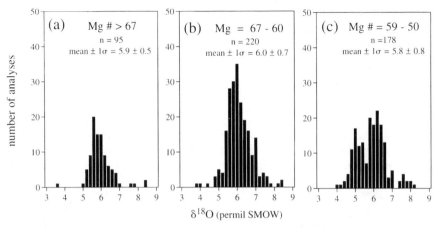

Fig. 36. δ^{18}O-frequency distributions of basalts as a function of Mg-number. (After Harmon and Hoefs 1995)

(Fig. 36). Primary, unmodified basalts can be recognized on the basis of their Mg-number. Two important points emerge from Fig. 36, which illustrates the variation in ^{18}O as a function of Mg-numbers. There is a 5‰ range from δ^{18}O of 3.6‰ to 8.7‰ for basalts with Mg-numbers 75–68 that can be considered primary mantle partial melts. However, both basalts with Mg > 75 that have accumulated olivine and more differentiated basalts with a lower Mg of 67–50 exhibit a similar large range in ^{18}O variation. Thus, no strong dependence of ^{18}O/^{16}O ratios is observed as a function of Mg-number. Second, although the form of the histogram differ for the three Mg-groups, their mean ^{18}O values are equivalent. Thus, overall the effects of differentiation and other petrogenetic processes on the O-isotope composition of basaltic magmas with Mg > 50 are small, probably no more than a few tenths per mill. Larger variations may be expected in particular situations where a differentiated basalt with a Mg < 50 has evolved through a combined process of assimilation-crystal fractionation from a primary parental basalt.

Table 18. O-isotope characteristics of basalts erupted in different tectonic settings

Tectonic setting	Number	^{18}O Range	Mean δ^{18}O±1σ
Oceanic basalts (all)	440	2.9 to 7.5	5.40 ± 0.75
MORB	127	5.2 to 6.4	5.73 ± 0.21
Ocean island basalt	148	4.6 to 7.5	5.48 ± 0.51
Iceland	104	2.9 to 6.2	4.50 ± 0.81
Ocean arc basalt	33	5.3 to 7.5	6.10 ± 1.10
Back-arc basalt	28	5.5 to 6.6	5.93 ± 0.26
Continental basalt (all)	303	4.3 to 11.4	6.36 ± 1.06
Cont. intraplate basalt	171	4.5 to 8.1	6.08 ± 0.66
Cont. arc basalt	82	4.8 to 7.7	6.24 ± 0.65
Cont. flood basalt	17	4.3 to 6.5	5.59 ± 0.64
Italy	33	6.3 to 11.4	8.47 ± 1.44

In summary, the substantial 5‰ variation should reflect substantial heterogeneity in the mantle source region. To elucidate this in more detail, Harmon and Hoefs (1995) have considered their database in terms of seven tectonic settings (Table 18) with Iceland and the intraplate potassic magmatism of Italy treated separately because of their special geotectonic situation.

As shown in Table 18, MORB has a remarkably uniform O-isotope composition of all basalt types (5.7 ± 0.2‰) and can be used as a reference against which basalts erupted in other tectonic settings can be compared. The O-isotope character of Ocean Island basalts (OIB) stands in contrast to that for MORB. The 3‰ variation extends to both higher and lower $\delta^{18}O$-values and cannot be explained in terms of the small ^{18}O fractionations that may occur during partial melting or during closed system fractional crystallization. Detailed studies of rocks from Hawaii, Iceland, and Afar have shown the widespread occurrence of low ^{18}O contents in OIBs, whereas Polynesian OIBs are characterized by higher $\delta^{18}O$ values than MORB. These features are a clear indication that the deep mantle which gives rise to plumes is substantially more heterogeneous than the depleted upper mantle. Continental basalts tend to be enriched in ^{18}O relative to oceanic basalts and exhibit considerably more variability in O-isotope composition, a feature attributed to interaction with ^{18}O-enriched continental crust during magma ascent. Systematic covariations between O-, Sr-, Nd-, and Pb-isotope ratios reflect the same intramantle end-member isotopic components (DMM, HIMU, EM-I, EM-II) deduced from radiogenic isotope considerations and imply that a common process is responsible for the upper mantle stable and radiogenic isotope heterogeneity.

3.3.6
Ocean Water/Basaltic Crust Interactions

Information about the O-isotope character of the oceanic crust comes from DSDP/ODP drilling sites and from the study of ophiolite complexes, which presumably represent pieces of ancient oceanic crust. Primary, unaltered oceanic crust should have $\delta^{18}O$-values close to MORB ($\delta^{18}O$: 5.7‰, Table 18). Two types of alteration can be distinguished within the oceanic lithosphere: at low temperatures weathering may markedly enrich the groundmass of basalts in ^{18}O, but not affect phenocrysts. The extent of this low-temperature alteration correlates with the water content: the higher the water content, the higher the $\delta^{18}O$-values. At temperatures in excess of about 300 °C, hydrothermal circulation beneath the mid-ocean ridges leads to a high-temperature water/rock interaction in which deeper parts of the oceanic crust become depleted in ^{18}O by 1–2‰. Similar findings have been reported from ophiolite complexes, the most cited example being that of Oman (Gregory and Taylor 1981). Maximum ^{18}O contents occur in the uppermost part of the pillow lava sequence and decrease through the sheeted dike complex. Below the base of the dike complex down to the Moho, $\delta^{18}O$-values are lower than mantle values by about 1–2‰.

Thus, separate levels of the oceanic crust are simultaneously enriched and depleted in ^{18}O relative to "normal" mantle values because of reaction with sea water at different temperatures. Muehlenbachs and Clayton (1976) and Gregory and

Taylor (1981) concluded that the ^{18}O enrichments are balanced by the ^{18}O depletions, which acts like a buffer for the oxygen isotope composition of ocean water.

3.3.7
Granitic Rocks

On the basis of their $^{18}O/^{16}O$ ratios, Taylor (1977, 1978) has subdivided granitic rocks into three groups: (1) normal ^{18}O-granitic rocks with $\delta^{18}O$-values between 6 and 10‰, (2) high ^{18}O granitic rocks with $\delta^{18}O$-values >10‰, and (3) low ^{18}O granitic rocks with $\delta^{18}O$-values <6‰. Although this is a somewhat arbitrary grouping, it nevertheless turns out to be a useful geochemical classification.

Many granitic plutonic rocks throughout the world have relatively uniform ^{18}O contents with $\delta^{18}O$-values between 6 and 10‰. Granitoids at the low ^{18}O end of the normal group have been described from oceanic island–arc areas where continental crust is absent (Chivas et al. 1982). Such plutons are considered to be entirely mantle-derived. Granites at the high end of the normal ^{18}O group may have formed by partial melting of crust that contained both a sedimentary and a volcanic fraction. It is interesting to note that many of the normal ^{18}O-granites are of Precambrian age and that metasediments of this age quite often have $\delta^{18}O$-values below 10‰ (Longstaffe and Schwarcz 1977).

Granitic rocks with $\delta^{18}O$-values higher than 10‰ require derivation from some type of ^{18}O-enriched sedimentary or metasedimentary protolith. For instance, such high $\delta^{18}O$-values are observed in many Hercynian granites of western Europe (Hoefs and Emmermann 1983), in Damaran granites of Africa (Haack et al. 1982), and in granites from the Himalayas of central Asia (Blattner et al. 1983). All these granites are easily attributed to anatexis within a heterogeneous crustal source, containing a large metasedimentary component.

Granitic rocks with $\delta^{18}O$-values lower than 6‰ cannot be derived by any known differentiation process from basaltic magmas. Excluding those low-^{18}O granites which have exchanged with ^{18}O-depleted meteoric-hydrothermal fluids under subsolidus conditions (see p. 95), a few primary low-^{18}O granitoids have been observed (Taylor 1987). These granites obviously inherited their ^{18}O depletion while still predominantly liquid, prior to cooling and crystallization. Such low-^{18}O magmas may be formed by remelting of hydrothermally altered country rocks or by large-scale assimilation of such material in a rift-zone tectonic setting.

3.4
Volatiles in Magmatic Systems

The isotope composition of magmatic volatiles can be deduced by analyses of glasses, volcanic gases, and hot springs. The main process which can cause isotope fractionation of volatile compounds is degassing. Magmatic rocks generally contain only a fraction of their original volatiles; the residual fraction will differ isotopically from the original volatiles. The other process which can alter the isotopic composition of magmatic volatiles is assimilation and contamination. The ultimate origin of volatiles in magmatic systems – whether juvenile in the sense that they originate from primary mantle degassing or recycled by subduction processes – is difficult to assess, but may be deduced in suitable cases.

3.4.1
Glasses

3.4.1.1
Hydrogen

Water dissolves in silicate melts and glasses in at least two distinct forms: water molecules and hydroxyl groups (Stolper 1982). Because the proportions of these two species change with total water content, temperature, and chemistry, the bulk partitioning of hydrogen isotopes between vapor and melt is a complex function of these variables. Dobson et al. (1989) determined the fractionation between water vapor and water dissolved in felsic glasses in the temperature range from 530 to 850 °C. Under these conditions, the total dissolved water contents of the glasses were below 0.2%, with all water present as hydroxyl groups. The measured hydrogen fractionation factors vary from 1.051 to 1.035 and are greater than those observed for most hydrous mineral–water systems, perhaps reflecting the strong hydrogen bonding of hydroxyl groups in glasses.

Hydrogen isotope and water content data for MORB, OIB, and BAB (back arc basalt) glasses have been determined by Kyser and O'Neil (1984), Poreda (1985), and Poreda et al. (1986). The range of δD-values for MORB glasses is from –90‰ to –40‰ and is indistinguishable from that reported on phlogopites and amphiboles from kimberlites and peridotites (see Fig. 33). Kyser and O'Neil (1984) demonstrated that D/H ratios and water contents in fresh submarine basalt glasses can be altered by (1) degassing, (2) addition of seawater at magmatic temperature, and (3) low-temperature hydration. Extrapolations to possible unaltered D/H ratios indicate the primary δD-value for most basalts of -80 ± 5‰.

The H-isotope composition of subaerial volcanic glasses can be modified by slow inward diffusion of meteoric water at low temperature (Kyser and O'Neil 1984). These authors observed that the outer rims of volcanic glasses can undergo low-temperature hydration by hydroxyl groups having δD-values of –100‰. In addition to low-temperature hydration, exsolution of water from a magma can deplete the magma in deuterium (Nabelek et al. 1983; Taylor et al. 1983). By contrast, a D enrichment would occur if either H_2 or CH_4 was lost from the melt because these gases are strongly depleted in D relative to the melt at magmatic temperatures (Richet et al. 1977).

3.4.1.2
Carbon

The solubility mechanism of CO_2 in basic melts is strongly dependent on melt structure and availability of suitable complexing cations such as Ca^{2+} and Mg^{2+}.

Reported $\delta^{13}C$-values for basaltic glass vary from -30 to about $-3‰$ and represent isotopically distinct carbon extracted at different temperatures by a stepwise heating (Pineau et al. 1976; Pineau and Javoy 1983; Des Marais and Moore 1984; Mattey et al. 1984). A "low-temperature" component of carbon is extractable below 600 °C, whereas a "high-temperature" fraction of carbon is liberated above 600 °C. There are two different interpretations regarding the origins of these two different types of carbon. While Pineau et al. (1976) and Pineau and Javoy (1983) consider that the whole range of carbon isotope variation observed represents primary dissolved carbon, which becomes ^{13}C depleted during multistage degassing of CO_2, Des Marais and Moore (1984) and Mattey et al. (1984) suggest that the "low-temperature" carbon originates from surface contamination. For MORB glasses, the "high-temperature" carbon falls in the range of typical mantle values. Island arc glasses have lower $\delta^{13}C$-values, which might be explained by mixing two different carbon compounds in the source regions: a MORB-like carbon and an organic carbon component from subducted pelagic sediments (Mattey et al. 1984).

3.4.1.3
Nitrogen

Nitrogen in basaltic glasses has been determined by Exley et al. (1987). At low temperatures <600 °C, glasses release small amounts of nitrogen with $\delta^{15}N$-values in the range of atmospheric nitrogen, suggesting surficial adsorption. At high temperatures around 1000 °C, $\delta^{15}N$-values vary from -4.5 to $15.5‰$. This large range may reflect partly mantle heterogeneities and partly varying degassing histories (Exley et al. 1987).

3.4.1.4
Sulfur

The behavior of sulfur in magmatic systems is particularly complex: sulfur can exist as both sulfate and sulfide species in four different forms: dissolved in the melt, as an immiscible sulfide melt, in a separate gas phase, and in various sulfide and sulfate minerals. MORB glasses and submarine Hawaiian basalts have a very narrow range in sulfur isotope composition, with $\delta^{34}S$-values clustering around zero (Sakai et al. 1982, 1984). In subaerial basalts, the variation of $\delta^{34}S$-values is larger and generally shifted towards positive values. One reason for this larger variation is the loss of a sulfur-bearing phase during magmatic degassing. The effect of this process on the sulfur isotope composition depends on the ratio of sulfate to sulfide in the magma, which is directly proportional to the fugacity of oxygen (Sakai et al. 1982). Arc volcanic rocks are particularly enriched in ^{34}S, with $\delta^{34}S$-values up to $+20‰$ reported by Ueda and Sakai (1984) and Harmon and Hoefs (1986). This overall enrichment is considered to be mainly a product of recycling of marine sulfate during subduction.

3.4.2
Volcanic Gases and Hot Springs

The chemical composition of volcanic gases is naturally variable and can be modified significantly during sampling due to differences in the temperature of collection and due to errors during the sampling process. While it is relatively simple to recognize and correct modifications from atmospheric contamination, the effects of natural contamination processes in the near-surface environment are a much more difficult issue to address. Thus, the identification of truely mantle derived gases except helium remains very problematic. In addition to assimilation/contamination processes, the degassing history can alter significantly the isotopic composition of magmatic volatiles.

3.4.2.1
Water

A long-standing geochemical problem is the source of the water in volcanic eruptions and geothermal systems: how much is derived from the magma itself and how much is recycled meteoric water? One of the principal and unequivocal conclusions drawn from stable isotope studies of fluids in volcanic hydrothermal systems is that most hot spring waters are meteoric waters derived from local precipitation (Craig et al. 1956; Craig 1966; Clayton et al. 1968; Clayton and Steiner 1975; Truesdell and Hulston 1980; and others).

Most hot spring waters have deuterium contents similar to those of local precipitation, but are usually enriched in ^{18}O as a result of isotopic exchange with the country rock at elevated temperatures. The magnitude of the oxygen isotope shift depends on the original $\delta^{18}O$-value of both water and rock, the mineralogy of the rock, the temperature, the water/rock ratio, and the time of contact.

Geothermal waters near ocean coasts may consist of mixtures of heated oceanic and local meteoric waters (Sakai and Matsubaya 1974). What is sometimes overlooked is the effects of boiling, e.g., loss of steam from a geothermal fluid can cause isotopic fractionations. Quantitative estimates of the effects of boiling on the isotopic composition of water can be made using known temperature-dependent fractionation coefficients and assumptions as to the extent to which the generated steam remains in contact with liquid water during the boiling process (Truesdell and Hulston 1980).

In recent years, there is increasing evidence that a magmatic water component cannot be excluded in some volcanic systems. As more and more data have become available from volcanoes around the world, especially from those at very high latitudes, Giggenbach (1992) demonstrated that "horizontal" ^{18}O shifts are actually the exception rather than the rule: oxygen isotope shifts are also accompanied by a deuterium shift (Fig. 37). Giggenbach (1992) argued that these waters all followed similar trends, corresponding to mixing of local ground waters with a water having a rather uniform isotopic composition with a $\delta^{18}O$-value of about 10‰ and a δD-value of about –20‰. He postulated the existence of a common magmatic component in andesite volcanoes having a δD of –20‰, which is much heavier than the gener-

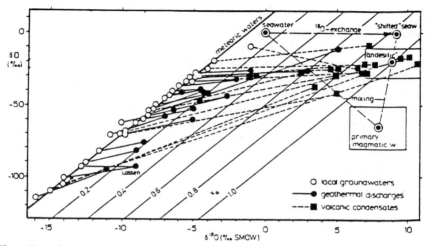

Fig. 37. Isotopic composition of thermal waters and of associated local groundwaters. Lines connect corresponding thermal waters to local groundwaters. (Giggenbach 1992; Reprinted from Earth and Planetary Science Letters, Vol. 113, Isotopic Shifts in Geothermal Waters at Convergent Plate Boundaries, p 497, © 1992, with kind permission from Elsevier Science – NL, Sara Burgerhartstraat 25, 1055 KV Amsterdam, The Netherlands)

ally assumed mantle water composition. The most likely source would be recycled seawater carried to the zones of arc magma generation by the subducted slab.

3.4.2.2
Carbon

CO_2 is the second most abundant gas species. In a survey of CO_2 emanations from tectonically active areas worldwide, Barnes et al. (1978) attributed $\delta^{13}C$-values between –8 and –4‰ to a mantle source. This is, however, problematical, because average crustal and mantle isotope compositions are more or less identical and surficial processes that can modify the carbon isotope composition are numerous. A more promising approach may be to analyze the ^{13}C content of CO_2 directly collected from magmas at high temperatures.

The volcano where gases have been collected and analyzed for the longest time is Kilauea in Hawaii, the database covering a period from about 1960 to 1985 (Gerlach and Thomas 1986; Gerlach and Taylor 1990). Gerlach and Taylor (1990) consider a $\delta^{13}C$-value of $-3.4 \pm 0.05‰$ to be the best estimate of the mean for the total summit gas emission of Kilauea. A two-stage degassing model was developed to explain this observation. This involved: (1) ascent and pressure equilibration in the summit magma chamber and (2) rapid, near surface decompression of summit-stored magma during ascent and eruption. The study demonstrated that the gas at the summit is a direct representation of the parental magma C-isotope ratio ($\delta^{13}C$: –3.4‰), whereas gases given off during East Rift Zone eruptions have a $\delta^{13}C$-value of –7.8‰, corresponding to a magma which had been affected by a degassing episode in a shallow magmatic system.

Very similar δ^{13}C-values have been observed for CO_2 released from MORB glass (-3.7‰, Javoy and Pineau 1991) and for CO_2 from the Iceland plume (-3.8‰, Poreda et al. 1992). A slightly heavier δ^{13}C-value of -3‰ for arc volcanoes from Indonesia (Poorter et al. 1991) is consistent with the contribution of small amounts of sediment being subducted. It is notable that the δ^{13}C-value of -3‰ estimated for the upper mantle on the basis of volcanic CO_2 is higher than the estimates of -5‰ on the basis of kimberlites and carbonatites.

Besides CO_2, methane has been reported in high-temperature hydrothermal vent fluids (Welhan 1988; Ishibashi et al. 1995). The origin of this methane is somewhat unclear, even in systems which are associated with ^{3}He anomalies. Whereas a nonbiogenic magmatic origin of methane has been assumed for the East Pacific Rise (Welhan 1988), a thermogenic origin has been proposed for the Okinawa trough (Ishibashi et al. 1995).

3.4.2.3
Sulfur

Elucidation of the origin of sulfur in volcanic systems is complicated by the fact that besides SO_2, H_2S, sulfate, and elemental sulfur can be present in appreciable amounts. The bulk sulfur isotope composition has to be calculated from mass balance considerations. The principal sulfur gas in equilibrium with basaltic melts at low pressure and high temperature is SO_2. With decreasing temperature and/or increasing water fugacity, H_2S becomes more stable. δ^{34}S-values of SO_2 sampled at very high temperatures provide the best estimate of the ^{34}S content of magmas (Taylor 1986). Sakai et al. (1982) reported δ^{34}S-values of 0.7‰–1‰ in the solfataric gases of Kilauea, which compare well with the δ^{34}S values of 0.9‰–2.6‰ for Mount Etna gases, observed by Allard (1983). SO_2 from volcanoes of andesitic and dacitic composition is more enriched in ^{34}S. This is especially pronounced in arc volcanoes from Indonesia, where Poorter et al. (1991) found a δ^{34}S-value of 5‰ for the bulk sulfur. Subducted oceanic crust may provide the ^{34}S-enriched sulfur to arc volcanoes.

In summary, stable isotope analysis (H,C,S) of volcanic gases and hot springs may permit inference about the isotopic composition of the mantle source. However, it must be kept in mind that numerous possibilities for contamination, assimilation, and gas phase isotopic fractionation, especially in the surficial environment, make such deductions problematic at best. In cases where it may possible to "look through" these secondary effects, small differences in H, C, and S isotope compositions of volcanic gases and hot springs might be characteristic of different geotectonic settings.

3.4.3
Isotope Thermometers in Geothermal Systems

Although there are many isotope exchange processes occurring within a geothermal fluid which have the potential to provide thermometric information, only a

Table 19. Isotope temperature and rates of exchange to establish equilibrium for the hydrothermal fluid at Wairakei, New Zealand. (Hulston 1977)

Element	Species	Isotope temperature	Rates of exchange
C	$^{13}CH_4 - {}^{12}CO_2$	350 °C	10^2-10^5 years
S	$H^{34}SO_4^- - H_2^{32}S$	350 °C	10^3 years
O	$S^{18}O_4^{2-} - H_2^{16}O$	280 °C	1 year
H	$H_2 - HDO$	260 °C	1–2 weeks
	Drill hole temperature	260 °C	

few have been generally applied, because of suitable exchange rates for achieving isotope equilibrium (Hulston 1977; Truesdell and Hulston 1980; Giggenbach 1982). In a single geothermal system, several different reservoirs may exist which generally increase in temperature with depth. The presence of such reservoirs at different temperatures may be indicated by different isotope thermometers equilibrating at different rates. The reaction rates of these exchange reactions determine whether they will equilibrate in deep geothermal reservoirs or if they will re-equilibrate in shallower reservoirs. Table 19 demonstrates that some isotope exchange reactions, such as methane–carbon dioxide or sulfate–hydrogen sulfide, do not equilibrate at temperatures below 350 °C, whereas others equilibrate so rapidly that only the temperature of collection is indicated.

3.5
Ore Deposits and Hydrothermal Systems

Stable isotopes have become an integral part of ore deposit studies. The determination of the light isotopes H, C, O, and S can provide information about the diverse origins of ore fluids, about temperatures of mineralization, and about physicochemical conditions of mineral deposition. In contrast to early views which assumed that almost all metal deposits owed their genesis to magmas, stable isotope investigations have convincingly demonstrated that ore formation has taken place in the Earth's near-surface environment by recycling processes of fluids, metals, sulfur, and carbon. More recent reviews of the application of stable isotopes to the genesis of ore deposits have been given by Ohmoto (1986) and Taylor (1987).

Inasmuch as water is the dominant constituent of ore-forming fluids, a knowledge of its origin is fundamental to any theory of ore genesis. There are two ways of determining δD- and δ18O-values of ore fluids: (1) by direct measurement of fluid inclusions contained within hydrothermal minerals or (2) by analysis of hydroxyl-bearing minerals and the calculation of the isotopic composition of fluids from known temperature-dependent mineral-water fractionations, assuming that minerals were precipitated from solutions under conditions of isotope equilibrium.

1. There are two different methods through which fluids and gases may be extracted from rocks: (1) thermal decrepitation by heating in vacuum and (2)

crushing and grinding in vacuum. Serious analytical difficulties may be associated with both techniques. The major disadvantage of the thermal decrepitation technique is that, although the amount of gas liberated is higher than by crushing, compounds present in the inclusions may exchange isotopically with each other and with the host mineral at the high temperatures necessary for decrepitation. Crushing in vacuum largely avoids isotope exchange processes; however, during crushing large new surfaces are created which easily absorb some of the liberated gases and which, in turn, might be associated with fractionation effects. Both techniques preclude separating the different generations of inclusions in a sample and, therefore, the results obtained represent an average isotopic composition of all generations of inclusions. Numerous studies have used the δD-value of the extracted water to deduce the origin of the hydrothermal fluid. However, without knowledge of the internal distribution of hydrogen in quartz, such a deduction can be misleading (Simon 1996). Hydrogen in quartz mainly occurs in two reservoirs: (1) in trapped fluid inclusions and (2) in small clusters of structurally bonded molecular water. Because of hydrogen isotope fractionation between the hydrothermal fluid and the structurally bonded water, the total hydrogen extracted from quartz does not necessarily reflect the original hydrogen isotope composition. This finding may explain why often δD-values from fluid inclusions tend to be lighter than δD-values from associated minerals (Simon 1996).

2. The indirect method of deducing the isotope composition of ore fluids is more frequently used, because it is technically easier. Uncertainties arise from several sources: uncertainty in the temperature of deposition and uncertainty in the equations for isotope fractionation factors. Another source of error is an imprecise knowledge also of the effects of fluid chemistry („salt effect") on mineral-water fractionation factors. Oxygen-bearing minerals crystallize during all stages of mineralization, the occurrence of hydrogen-bearing minerals being restricted in most ore deposits. Examples of hydroxyl-bearing minerals include biotite at high temperatures (in porphyry copper deposits), chlorite and sericite at temperatures around 300 °C, and kaolinite at around 200 °C. The mineral alunite, and its iron equivalent jarosite, are a special case. Alunite $(KAl_3(SO_4)_2(OH)_6)$ contains four stable isotope sites and both the sulfate and hydroxyl anionic groups may provide information on fluid source and condition of formation. Alunite forms under highly acidic oxidizing conditions and is characterized by the assemblage alunite+kaolinite+quartz±pyrite. Stable isotope data of alunite in combination with associated sulfides and kaolinite permit recognition of environments and temperatures of formation (Rye et al. 1992).

3.5.1
Origin of Ore Fluids

Ore fluids may be generated in a variety of ways. The principal types include seawater, meteoric waters, and juvenile water all of which have a strictly defined isotopic composition. All other possible types of ore fluids such as formation, meta-

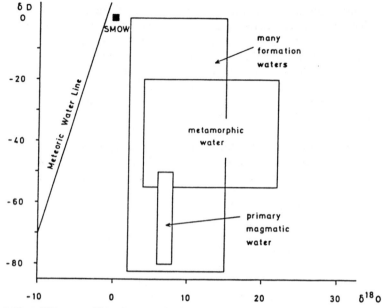

Fig. 38. Plot of δD versus δ^{18}O of waters of different origins

morphic, and magmatic waters can be considered recycled derivatives or mixtures of one or more of the three reference waters (see Fig. 38).

3.5.1.1
Seawater

The isotopic composition of present-day ocean water is more or less constant, with δ-values very near zero. The isotopic composition of ancient ocean water is less well constrained (see Sect. 3.8), but still should not be far removed from zero. Many volcanogenic massive sulfide deposits are formed in submarine environments from heated oceanic waters. This concept gains support from the recently observed hydrothermal systems at ocean ridges, where measured isotopic compositions of fluids are only slightly modified relative to 0‰.

    Bowers and Taylor (1985) have modeled the isotopic composition of an evolving seawater hydrothermal system. At low temperatures, the δ^{18}O-value of the fluid decreases relative to ocean water because the alteration products in the oceanic crust are ^{18}O rich. At around 250 °C, the solution returns to its initial seawater isotopic character. Further reaction with basalt at 350 °C increases the isotopically modified seawater to 2‰. The δD-value of the solution increases slightly at all temperatures because mineral-water fractionations are nearly all less than zero. At 350 °C, the δD-value of the solution is 2.5‰. The best-documented example of the role of ocean water during ore deposition is the Kuroko-type deposits (see the extensive monograph by Ohmoto and Skinner 1983).

3.5.1.2
Meteoric Waters

The δD- and $\delta^{18}O$-values of meteoric waters vary together linearly on a global scale and are dependent on geographic location. D/H and $^{18}O/^{16}O$ ratios become increasingly lower with increasing distance from the equatorial region to the polar regions, from coastal to inland regions, and with elevation. The meteoric water line of the geological past may have moved somewhat but should remain parallel to the present-day meteoric water line ($\delta D = 8\delta^{18}O + 10$). Heated meteoric waters are a major constituent of ore-forming fluids in many ore deposits and may become dominant in the latest stages of ore deposition. The isotopic variations observed for several Tertiary North American deposits are systematic with latitude and, hence, palaeo-meteoric water composition (Sheppard et al. 1971). The ore-forming fluid has commonly been shifted in O-isotope composition from its meteoric $\delta^{18}O$-value to higher ^{18}O contents through water-rock interaction.

3.5.1.3
Juvenile Water

The concept of juvenile water has influenced tremendously early discussions about ore genesis. The terms "juvenile water" and "magmatic water" have been used synonymously sometimes, but they are not exactly the same. Juvenile water originates from degassing of the mantle and has never existed as surface water. Magmatic water is a nongenetic term and simply means a water that has equilibrated with a magma.

It is difficult to prove that juvenile water has ever been sampled. One way to search for juvenile water is by analyzing hydroxyl-bearing minerals of mantle origin (Sheppard and Epstein 1970). The estimated isotopic composition of juvenile water from such an approach is δD: -60 ± 20‰ and $\delta^{18}O$: $+6 \pm 1$‰ (Ohmoto 1986).

3.5.1.4
Magmatic Water

Despite the close association of intrusions with many ore deposits, there is still debate about the extent to which magmas contribute water and metals to ore-forming fluids. Many early studies of the stable isotope composition of hydrothermal minerals indicated a dominance of meteoric water (Taylor 1974), more recent studies showing that magmatic fluids are commonly present, but that their isotopic signatures may be masked or erased during later events such as the influx of meteoric waters (Rye 1993; Hedenquist and Lowenstern 1994).

The δD-value of magmatic water changes progressively during degassing, resulting in a positive correlation between δD and the residual water content of an igneous body. The variation in D contents depends on whether the degassing is dominantly an open- or closed-system process (Taylor 1986). Thus, late-formed hydoxyl-bearing minerals represent the isotopic composition of a degassed melt rather than that of the initial magmatic water. The δD of most of the water exsolved from many felsic melts is in the range of -60 to $-30‰$, whereas the associated magmatic rocks may be significantly depleted in D.

The calculated range of isotopic composition for magmatic waters is commonly 6–10‰ for $\delta^{18}O$-values and -50–80‰ for δD-values. Unpublished experimental data cited in Ohmoto (1986) suggest some small pressure and compositional effects on the mineral-water isotope fractionations. Furthermore, magmatic fluids may change their isotopic composition during cooling through isotope exchange with country rocks and mixing with fluids entrained within the country rocks. Thus, the participation of a magmatic water component during an ore-forming process is generally not easily detected.

3.5.1.5
Metamorphic Water

Metamorphic water is defined as water associated with metamorphic rocks during metamorphism. Thus, it is a descriptive, nongenetic term and may include waters of different origins. In a narrower sense, metamorphic water refers to the fluids generated by dehydration of minerals during metamorphism. The isotopic composition of metamorphic water may be highly variable, depending on the respective rock types and their history of fluid/rock interaction. A wide range of $\delta^{18}O$-values (5–25‰) and δD-values (-70–20‰) is generally attributed to metamorphic waters (Taylor 1974).

3.5.1.6
Formation Waters

The changes in the D and ^{18}O contents of pore fluids depend on the type of initial fluid (ocean water, meteoric water), the temperature, and the lithology of rocks with which the fluids are or have been associated. Generally, formation waters with the lowest temperature and salinity have the lowest δD- and $\delta^{18}O$-values, approaching those of meteoric waters. Brines of the highest salinities are generally more restricted in isotopic composition. It is still an unanswered question whether meteoric water was the only source of these brines, with the final isotope composition produced by reactions between meteoric water and sediments, or whether the brines represent mixtures of fossil ocean water and meteoric water (see discussion on p.114).

3.5.2
Wall-Rock Alteration

Information about the origin and genesis of ore deposits can also be obtained by analyzing the alteration products in wall-rocks. Hydrogen and oxygen isotope zonation in wall-rocks around hydrothermal systems can be used to define the size and the conduit zones of a hydrothermal system. The fossil conduit is a zone of large water fluxes, generally causing a strong alteration and depletion in $\delta^{18}O$-values. Thus, fossil hydrothermal conduits can be outlined by following the zones of ^{18}O depletion. Oxygen isotope data are especially valuable in rock types which do not show diagnostic alteration of mineral assemblages as well as those in which the assemblages have been obliterated by subsequent metamorphism (Beaty and Taylor 1982; Green et al. 1983). Criss et al. (1985, 1991) found excellent spatial correlations between low $\delta^{18}O$-values and economic mineralization. Thus, zones having anomalously low ^{18}O contents may be a useful guide for exploration of hydrothermal ore deposits.

3.5.3
Fossil Hydrothermal Systems

Mainly through the work of H.P.Taylor and coworkers, it has become well established that many epizonal igneous intrusions have interacted with meteoric groundwaters on a very large scale. The interaction and transport of large amounts of meteoric water through hot igneous rocks produces a depletion in ^{18}O in the igneous rocks by up to 10–15‰ and a corresponding shift in the ^{18}O content of the water. About 60 such systems have been observed to date (Criss and Taylor 1986): they exhibit great variations in size from relatively small intrusions (<100 km^2) to large plutonic complexes (>1000 km^2). Among the best-documented examples are the Skaergaard intrusion in Greenland, the Tertiary intrusions of the Scottish Hebrides, and the Tertiary epizonal intrusions of the northwestern United States and southern British Columbia, where 5% of the land surface has been altered by meteoric hydrothermal water (Criss et al. 1991).

The best-studied example of a hydrothermal system associated with a gabbro is that of the Skaergaard intrusion (Taylor and Forester 1979; Norton and Taylor 1979). The latter authors carried out a computer simulation of the Skaergaard hydrothermal system and found a good match between calculated and measured $\delta^{18}O$-values. They further demonstrated that most of the sub-solidus hydrothermal exchange took place at very high temperatures (400–800 °C), which is compatible with the general absence of hydrous alteration products in the mineral assemblages and with the presence of clinopyroxene.

In granitic hydrothermal systems, temperatures of alteration are significantly lower because of differences in the intrusion temperatures. The most conspicuous petrographic changes are chloritization of mafic minerals, particularly of biotite, and a major increase in the turbidity of feldspars. Isotopically, large non-equilibrium quartz–feldspar fractionations are typical. Steep linear trajectories on plots of $\delta^{18}O_{(feldspar)}$ versus $\delta^{18}O_{(quartz)}$ are a characteristic feature of these hydrothermally altered rocks (see Fig. 24). The trajectories result from the fact that feldspar exchanges ^{18}O with hydrothermal fluids much faster than coexisting quartz and from the fact that the fluids entering the rock system have $\delta^{18}O$-values which are out of equilibrium with the mineral assemblage. The process seldom goes to completion, so the final mineral assemblage is in isotope disequilibrium, which is the most obvious fingerprint of the hydrothermal event.

Taylor (1988) distinguished three types of fossil hydrothermal systems on the basis of varying water/rock ratios, temperatures, and the length of time that fluid/rock interaction proceeds:

Type I: Epizonal systems with a wide variation in whole rock ^{18}O contents and extreme oxygen isotope disequilibrium among coexisting minerals. These systems typically have temperatures between 200 and 600 °C and lifetimes <10^6 years.

Type II: Deeper-seated and/or longer-lived systems, also with a wide spectrum of whole rock $^{18}O/^{16}O$ ratios, but with equilibrated $^{18}O/^{16}O$ ratios among coexisting minerals. Temperatures are between 400 and 700 °C and lifetimes >10^6 years.

Type III: Equilibrated systems with a relatively uniform oxygen isotope composition in all lithologies. These systems require a large water/rock ratio, temperatures between 500 and 800 °C, and times around 5×10^6 years.

These types are not mutually exclusive; type III systems for example may have been subjected to type I or type II conditions at an earlier stage of their hydrothermal history.

3.5.4
Hydrothermal Carbonates

The measured $\delta^{13}C$- and $\delta^{18}O$-values of carbonates can be used to estimate the carbon and oxygen isotope composition of the fluid in the same way as has been dis-

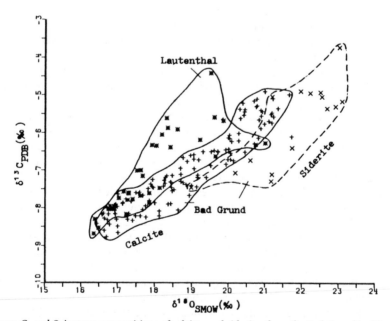

Fig. 39. C- and O-isotope compositions of calcites and siderites from the Bad Grund and Lautenthal deposits, Harz, Germany. (After Zheng and Hoefs 1993; Reprinted from Müller and Lüders (eds), Monograph Series on Mineral Deposits, Vol. 30, © 1993, with kind permission from Gebrüder Borntraeger Verlagsbuchhandlung, Johannesstraße 3A, 70176 Stuttgart, Germany)

cussed before for oxygen and hydrogen. The isotopic composition of carbon and oxygen in any carbonate precipitated in isotopic equilibrium with a fluid depends on the isotopic composition of carbon and oxygen in the fluid, the temperature of formation, and the relative proportions of dissolved carbon species (CO_2, H_2CO_3, HCO_3^-, and/or CO_3^{2-}). To determine carbonate speciation, pH and temperature must be known; however, in most geological fluids with temperatures above about 100 °C, the content of HCO_3^- and CO_3^{2-} is negligible compared to that of CO_2 and H_2CO_3.

Experimental investigations have shown that the solubility of carbonate increases with decreasing temperature. Thus, carbonate cannot be precipitated from a hydrothermal fluid due to simple cooling in a closed system. Instead, an open system is required in which processes such as CO_2 degassing, fluid-rock interaction, or fluid mixing can cause the precipitation of carbonate. These processes result in correlation trends in $\delta^{13}C$ vs. $\delta^{18}O$ space for hydrothermal carbonates as often observed in nature and theoretically modeled by Zheng and Hoefs (1993).

Figure 39 presents $\delta^{13}C$- and $\delta^{18}O$-values of hydrothermal carbonates from the Pb-Zn deposits of Bad Grund and Lautenthal, Germany. The positive correlation between $^{13}C/^{12}C$ and $^{18}O/^{16}O$ ratios can be explained either by calcite precipitation due to the mixing of two fluids with different NaCl concentrations or by calcite precipitation from a H_2CO_3-dominant fluid due to a temperature effect coupled with either CO_2 degassing or with fluid-rock interaction.

3.5.5
Sulfur Isotope Composition of Ore Deposits

A huge literature exists about the sulfur isotope composition in hydrothermal ore deposits. Some of this information has been discussed in earlier editions and, therefore, is not repeated here. Out of the numerous papers on the subject the reader is referred to comprehensive reviews by Rye and Ohmoto (1974), Ohmoto and Rye (1979), Nielsen (1985), Ohmoto (1986), and Taylor (1987). The basic principles to be followed in the interpretation of $\delta^{34}S$-values in sulfidic ores were elucidated by Sakai (1968), and were subsequently extended by Ohmoto (1972).

The isotopic composition of a hydrothermal sulfide is determined by a number of factors such as: (1) temperature of deposition, (2) isotopic composition of the hydrothermal fluid from which the mineral is deposited, (3) chemical composition of the dissolved element species including pH and fO_2 at the time of mineralization, and (4) relative amount of the mineral deposited from the fluid.

3.5.5.1
Importance of Changing fO2 and pH

First, consider the effect of pH increase due to the reaction of an acidic fluid with carbonatic host rocks. At pH = 5, practically all of the dissolved sulfur is undissociated H_2S, whereas at pH = 9 the dissolved sulfide is almost entirely dissociated. Since H_2S concentrates ^{34}S relative to dissolved sulfide ion, an increase in pH leads directly to an increase in the $\delta^{34}S$ of precipitated sulfides.

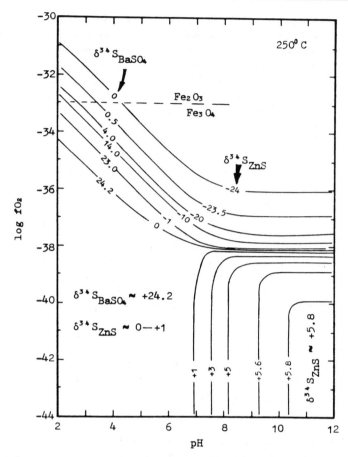

Fig. 40. Influence of fO_2 and pH on the sulfur isotope composition of sphalerite and barite at 250 °C and $\delta^{34}S_{\Sigma S}=0$. (Modified after Ohmoto 1972; Reprinted from Müller and Lüders (eds), Monograph Series on Mineral Deposits, Vol. 30, © 1993, with kind permission from Gebrüder Borntraeger Verlagsbuchhandlung, Johannesstraße 3A, 70176 Stuttgart, Germany) Reprinted from Müller and Lüders (eds), Monograph Series on Mineral Deposits, Vol. 30, © 1993, with kind permission from Gebrüder Borntraeger Verlagsbuchhandlung, Johannesstraße 3A, 70176 Stuttgart, Germany)

An increase in oxygen fugacities has a much stronger effect on the $\delta^{34}S$-values than a pH change, because of the large isotope fractionation between sulfate and sulfide. Figure 40 shows an example of the effect of pH and fO_2 variation on the sulfur isotope compositions of sphalerite and barite in a closed system at 250 °C with $\delta^{34}S_{\Sigma S}=0‰$. The curves are $\delta^{34}S$ contours, which indicate the sulfur isotope compositions of the minerals in equilibrium with the solution. Sphalerite $\delta^{34}S$-values can range from −24 to +5.8‰ and those for barite from about 0‰ to 24.2‰ within geologically reasonable limits of pH and fO_2. In the low fO_2 and pH region, sulfide ^{34}S contents can be similar to $\delta^{34}S_{\Sigma S}$ and can be rather insensitive to pH and fO_2 changes. In the region of high fO_2 values where the proportion of sulfate species becomes significant, mineral $\delta^{34}S$-values can be greatly different from

$\delta^{34}S_{\Sigma S}$ and small changes in pH or fO_2 may result in large changes in the sulfur isotope composition of either sulfide or sulfate. Such a change must, however, be balanced by a significant change in the ratio of sulfate to sulfide.

3.5.5.2
Reservoir Effect

When sulfide minerals precipitate from fluids, sulfur isotope fractionations occur
between dissolved sulfur species and sulfide minerals. The removal of sulfide
from the fluids in a closed system can cause a measurable fractionation on H_2S remaining in the fluids. This is called the reservoir effect.

An illustration of how the sulfur isotope compositions of precipitating sulfides are affected by the fraction of sulfur remaining in ore fluids is shown in
Fig. 41. Suppose that the sulfur in an ore fluid is dominated by H_2S, the sulfur isotope composition of the fluid system is 0‰, and the temperature 150 °C. Under
these conditions galena can have $\delta^{34}S$-values about –3‰ at the very beginning of
precipitation but more than 6‰ at the end of precipitation. Conversely, pyrite
deposited under the same conditions can decrease from initial $\delta^{34}S$-values of
about 2 to below –4‰. The reservoir effect is not significant for sphalerite and
chalcopyrite, because these two sulfides have very small fractionation factors
relative to H_2S.

Fig. 41. Influence of the fraction of sulfur remaining in
hydrothermal solution on the
S-isotope composition of sulfides during Rayleigh-type
precipitation at 150 °C with
$\delta^{34}S_{\Sigma S}$=0. (After Zheng and
Hoefs 1993;

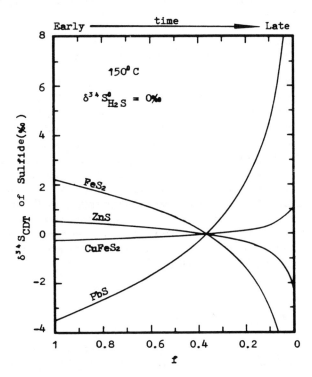

Table 20. Sulfur isotope compositions of vent fluids and chimneys. (Bowers 1989)

Vent	$\delta^{34}S_{fluid}$	$\delta^{34}S_{chimney}$
East Pacific Rise 21 °N		
OBS	1.3–1.5	0.9–6.2
SW	2.7–5.5	1.4–4.0
NGS	3.4	1.5–2.9
HG	2.3–3.2	1.5–2.6
East Pacific Rise 11–13 °N		
1	–	1.7–4.3
2	4.7	4.2–4.5
3	2.3–3.5	3.7–4.8
4	4.6	4.1–4.7
5	4.7–4.9	3.7–4.9
6	4.1–5.2	3.7–4.2
Southern Juan de Fuca Ridge		
Plume	4.2–7.3	1.6–5.7
1	4.0–6.4	2.1–4.0
3	7.2–7.4	4.3–5.0

In summary, interpretation of the distribution of $\delta^{34}S$-values relies on information about the source of sulfur and on a knowledge of the mineral parageneses that constrain the ambient temperature, E_H and pH. If the oxidation state of the fluid is below the sulfate/H_2S boundary, then the $^{34}S/^{32}S$ ratios of sulfides will be insensitive to redox shifts.

In the following section different classes of ore deposits are discussed.

3.5.5.3
Recent Sulfide Deposits at Mid-Ocean Ridges

Numerous sulfide deposits have been discovered in recent years on the seafloor along the East Pacific Rise, Juan de Fuca Ridge, Explorer Ridge, and Mid-Atlantic Ridge. These deposits are formed from hydrothermal solutions which result from the interaction of circulating seawater with oceanic crust. The role of sulfur in these vents is complex and often obscured by its multiple redox states and by uncertainties in the degree of equilibration. Studies by Styrt et al. (1981), Arnold and Sheppard (1981), Skirrow and Coleman (1982), Kerridge et al. (1983), Zierenberg et al. (1984), and others have shown that the sulfur in these deposits is enriched in ^{34}S relative to a mantle source, implying the addition of sulfide derived from seawater (see Table 20). Bowers (1989) modeled the effects of variable S-contributions from basalt and from inorganic reduction of oceanic sulfate and demonstrated that the overwhelming influence on the sulfur isotope composition of vent fluids is basaltic sulfur.

3.5.5.4
Magmatic Ore Deposits

Magmatic deposits are sulfides which precipitate from mafic silicate melts rather than hydrothermal fluids. Some economically important Ni–Cu deposits such as Sudbury and Noril'sk belong to this category. The characteristic features of these deposits are relatively large deviations in δ^{34}S-values from the presumed mantle melt value near zero. This feature is generally attributed to assimilation of sulfur from the wall rocks, whether present as sulfide or as sulfate. For a more detailed discussion of this type of deposit see *Reviews in Economic Geology 4* (1989).

3.5.5.5
Magmatic Hydrothermal Deposits

This group of deposit is closely associated in space and time with magmatic intrusions that were emplaced at relatively shallow depths. They have been developed in hydrothermal systems driven by the cooling of magma (e.g., porphyry-type deposits and skarns). From δD and δ^{18}O measurements, it has been concluded that porphyry copper deposits show the clearest affinity of a magmatic water imprint (Taylor 1974).

The majority of δ^{34}S-values of sulfides fall between –3 and 1‰ and of sulfates between 8 and 15‰ (Field and Gustafson 1976; Shelton and Rye 1982).

Calculated sulfate–sulfide temperatures, for conditions of complete isotope equilibrium, are typically between 450 and 600 °C and agree well with temperatures estimated from other methods. Thus, the sulfur isotope data and temperatures support the magmatic origin of the sulfur in porphyry deposits.

3.5.5.6
Base and Precious Metal Vein Deposits

A wide spectrum of ore deposits of a different nature occurs in this category. Rye and Ohmoto (1974) have demonstrated the difficulty in interpreting the genesis of such ores from their sulfur isotope ratios. The only meaningful classification seems to be related to the temperatures of ore deposition. Typical temperatures of mineralization range from 150 to 350 °C with variable salinities.

Individual deposits often reveal that more than one type of fluid, with distinct O, H, and sometimes S and C isotopic signatures, was involved in the formation of a single ore deposit. One of the fluids involved often appears to be of meteoric origin. In many deposits different fluids were alternatively discharged into the vein system and promoted the precipitation of a specific suite of minerals, such as one fluid precipitating sulfides and another precipitating carbonates (Ohmoto 1986).

Vein deposits, due to their vertical extent, frequently exhibit a pronounced isotope zoning. Differences in sulfur isotope ratios may result from temperature differences and from an increase in fO_2 and/or pH toward the surface.

3.5.5.7
Volcanic-Associated Massive Sulfide Deposits

This group of deposits is characterized by massive Cu–Pb–Zn–Fe sulfide ores associated with submarine volcanic rocks. They appear to have been formed near the seafloor by submarine hot springs at temperatures of 150–350 °C and may be regarded as analogues of the recent sulfide mineralizations at the oceanic ridges. Massive sulfide deposits have δ^{34}S-values typically between zero and the δ-value of contemporaneous oceanic sulfate, whereas the sulfate has δ-values similar to or higher than contemporaneous seawater. According to Ohmoto et al. (1983), the ore-forming fluid is evolved seawater fixed as disseminated anhydrite and then reduced by ferrous iron and organic carbon in the rocks.

3.5.5.8
Shale/Carbonate-Hosted Massive Sulfide Deposits

Like volcanic massive sulfide deposits, this group has formed on the seafloor or in unconsolidated marine sediments. Its members differ from volcanogenic massive deposits in that the dominant host-rock lithologies are marine shales and carbonates, the associated igneous activity is minor or negligible, and water depths seem to be considerably less than the >2000 m proposed for most volcanogenic deposits. The total range of sulfide δ^{34}S-values is much larger than the range observed in volcanogenic massive sulfide deposits.

Sulfides are fine grained and texturally complex, containing multiple generations of minerals. Two different origins of sulfur can be envisaged: biogenic and hydrothermal. Mineral separation methods cannot ensure that mineral separates contain only one type of sulfur. Therefore, conventional techniques cannot answer questions such as: is most of the sulfur produced by bacterial reduction of seawater or is it inorganically acquired and hydrothermally introduced together with the metals? In situ ion microprobe techniques allow isotope analysis on a scale as small as 20 µm. Studies by Eldridge et al. (1988, 1993) have revealed extremely large variations in a distance of millimeters with gross disequilibrium between base metal sulfides and overgrown pyrites. Thus, the mean δ^{34}S of such deposits is not particularly diagnostic of its origin.

3.5.5.9
Mississippi Valley Type Deposits

The Mississippi Valley Type (MVT) deposits are epigenetic Zn–Pb deposits which mainly occur in carbonates. In contrast to the shale/carbonate-hosted massive sulfide deposits, the Mississippi Valley Type deposits appear to have been formed in continental settings (Ohmoto 1986).

Characteristics often ascribed to MVT deposits include temperatures generally <200 °C and deposition from externally derived fluids, possibly basinal brines. The δ^{34}S range is large between deposits but relatively small within deposits. Regional variations with a northward δ^{34}S increase in the midcontinent districts and a westward increase in the Appalachian districts suggest thermal reduction of evaporite sulfate as the main sulfur source (Jones et al. 1996). Using SHRIMP ion microprobe measurements, McKibben and Eldridge (1995) observed large δ^{34}S variations from –10 to +25‰ within and among intergrown sulfides. This strong heterogeneity indicates a complex history of mineral deposition including mixing between two or more isotopically distinct fluids.

3.5.5.10
Biogenic Deposits

The discrimination between bacterial sulfate and thermal sulfate reduction in ore deposits on the basis of δ^{34}S-values is rather complex. The best criterion to distinguish between both types is the internal spread of δ-values. If individual sulfide grains at a distance of only a few millimeters exhibit large and nonsystematic differences in δ^{34}S-values, then it seems reasonable to assume an origin involving bacterial sulfate reduction. Irregular variations in ^{34}S contents are attributed to bacteria growing in reducing microenvironments around individual particles of organic matter. In contrast, thermal sulfate reduction requires higher temperatures supplied by external fluids, which is not consistent with the closed system environment of bacterial reduction.

Two types of deposits, where the internal S-isotope variations fit the expected scheme of bacterial reduction, but where the biogenic nature was already known earlier from other geological observations, are the "sandstone-type" uranium mineralization in the Colorado Plateau (Warren 1972) and the Kupferschiefer in central Europe (Marowsky 1969).

3.5.5.11
Metamorphosed Deposits

It is generally assumed that metamorphism reduces the isotopic variations in a sulfide ore deposit. Recrystallization, liberation of sulfur in fluid and vapor phases, such as the breakdown of pyrite into pyrrhotite and sulfur, and diffusion at elevated temperatures should tend to reduce initial isotopic heterogeneities.

Studies of regionally metamorphosed sulfide deposits (Seccombe et al. 1985; Skauli et al. 1992) indicate, however, little evidence of homogenization on the deposit scale. Significant changes may take place in certain restricted parts of the deposit as a result of special local conditions, controlled by factors such as fluid flow regimes and tectonics. Thus, a very limited degree of homogenization takes place during metamorphism (Cook and Hoefs, in press). The extent of this is obscured by primary distribution and zonation patterns.

3.6
Hydrosphere

First, some definitions concerning water of different origin are given. The term "meteoric" applies to water that has been a part of the meteorological cycle, and participated in processes such as evaporation, condensation, and precipitation.

All continental surface waters, such as rivers, lakes, and glaciers, fall into this general category. Because meteoric water may seep into the underlying rock strata, it will also be found at various depths within the lithosphere. The ocean, although it continuously receives the continental run-off of meteoric waters as well as rain, is not regarded as being meteoric in nature. Connate water is water which has been trapped in sediments at the time of burial. Formation water is present in rocks immediately before drilling and may be a useful nongenetic term for waters of unknown origin and age.

3.6.1
Meteoric Water

When water evaporates from the surface of the ocean, the water vapor is enriched in H and ^{16}O because $H_2^{16}O$ has a higher vapor pressure than HDO and $H_2^{18}O$ (Table 1). Under equilibrium conditions at 25 °C, the fractionation factors for evaporating water are 1.0092 for ^{18}O and 1.074 for D (Craig and Gordon 1965). However, under natural conditions, the actual isotopic composition of water is more negative than the predicted equilibrium values, due to kinetic effects (Craig and Gordon 1965). Vapor leaving the surface of the ocean cools as it rises and rain forms when the dew point is reached. During removal of rain from a moist air mass, the residual vapor is continuously depleted in the heavy isotopes, because the rain leaving the system is enriched in ^{18}O and D. If the air mass moves poleward and becomes cooler, additional rain formed will contain less ^{18}O than the initial rain. This relationship is schematically shown in Fig. 42. The isotope composition of mean worldwide precipitation is estimated to be $\delta D = -22‰$ and $\delta^{18}O = -4‰$ (Craig and Gordon 1965).

The International Atomic Energy Agency (IAEA) has conducted a worldwide survey of the isotope composition of monthly precipitation for more than 35 years. The global distribution of D and ^{18}O in rain has been monitored since 1961 through a network of stations (Yurtsever 1975). Up to 1993, approximately

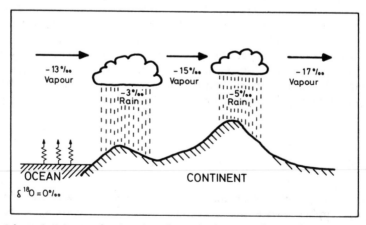

Fig. 42. Schematic O-isotope fractionation of water in the atmosphere. (After Siegenthaler 1979)

Fig. 43. Average δD-value of the annual precipitation from oceanic islands as a function of the average amount of annual rainfall. The island stations are distant from continents, within 30° of the equator, and at elevations less than 120 m. (After Lawrence and White 1991)

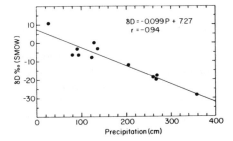

180 000 sets of isotope and meteorological data were accumulated by the IAEA (Rozanski et al. 1993). From this extensive database it can be deduced how geographic factors influence the isotopic composition of precipitation.

The first detailed evaluation of the equilibrium and nonequilibrium factors that determine the isotopic composition of precipitation was published by Dansgaard (1964). He demonstrated that the observed geographic distribution in isotope composition is related to a number of environmental parameters that characterize a given sampling site, such as latitude, altitude, distance to the coast, amount of precipitation, and surface air temperature. Out of these, two factors are of special significance: temperature and the amount of precipitation. The best temperature correlation is observed in continental regions nearer to the poles, whereas the correlation with amount of rainfall is most pronounced in tropical regions, shown in Fig. 43 (Lawrence and White 1991). The apparent link between local surface air temperature and the isotope composition of precipitation is of special interest mainly because of the potential importance of stable isotopes as palaeoclimatic indicators. The amount effect is ascribed to deep cooling of the air in heavy rainfall with only slight enrichments possible in later evaporation.

The theoretical approaches to explain isotope variations in meteoric waters evolved from the "isolated air mass" models, which are based on Rayleigh condensation, with immediate removal of precipitation or with a part of the condensate being kept in the cloud during the rain-out process. Isotope studies of individual rain events have revealed that successive portions of single events may vary drastically (Rindsberger et al. 1990; Schirmer 1995). Quite often the pattern is "V-shaped," a sharp decrease of δ-values usually being observed at the beginning of a storm with a minimum somewhere in the middle of the event. The most depleted isotope values correspond usually to the period of most intense rain. It has also been observed that convective clouds produce precipitation with higher δ-values than stratiform clouds. Thus, the isotope composition of precipitation from a given rain event depends on meteorological history of the air mass in which the precipitation is produced and the type of cloud through which it falls. Liquid precipitation (rain) and solid precipitation (snow, hail) may differ in their isotope composition insofar as rain drops may undergo evaporation and isotope exchange with atmospheric vapor on their descent to the surface. By analyzing hailstones, discrete meteorological events can be studied because hailstones keep a record on the internal structure of a cloud. Jouzel et al. (1975) concluded that hailstones grow during a succession of upward and downward movements in a cloud.

3.6.1.1
δD–δ¹⁸O Relationship

In all processes concerning evaporation and condensation, hydrogen isotopes are fractionated in proportion to oxygen isotopes, because a corresponding difference in vapor pressures exists between H_2O and HDO in one case and $H_2^{16}O$ and $H_2^{18}O$ in the other. Therefore, hydrogen and oxygen isotope distributions are correlated in meteoric waters. Craig (1961) first defined the following relationship:

$$\delta D = 8 \, \delta^{18}O + 10$$

which is generally known as the "Global Meteoric Water Line."

Later, Dansgaard (1964) introduced the concept of "deuterium excess," d defined as $d = \delta D - 8 \, \delta^{18}O$. Neither the numerical coefficient, 8, nor the deuterium excess, d, are really constant, both depending on local climatic processes. The long-term arithmetic mean for all analyzed stations of the IAEA network (Rozanski et al. 1993) is:

$$\delta D = (8.17 \pm 0.06) \, \delta^{18}O + (10.35 \pm 0.65) \quad r^2 = 0.99, \, n = 206$$

Relatively large deviations from the general equation are evident when monthly data for invidual stations are considered (Table 20). In an extreme situation, represented by the St. Helena station, a very poor correlation between D and ¹⁸O exists. At this station, it appears that all precipitation comes from nearby sources and represents the first stage of the rain-out process. Thus, the generally weaker correlations for the marine stations (Table 21) may reflect varying contributions of air masses with different source characteristics and a low degree of rain-out.

Table 21. Variations in the numerical constant and the deuterium excess for selected stations of the IAEA global network. (Rozanski et al. 1993)

Station	Numerical constant	Deuterium excess	r^2
Continental and coastal stations			
Vienna	7.07	−1.38	0.961
Ottawa	7.44	+5.01	0.973
Addis Ababa	6.95	+11.51	0.918
Bet Dagan,Israel	5.48	+6.87	0.695
Izobamba(Ecuador)	8.01	+10.09	0.984
Tokyo	6.87	+4.70	0.835
Marine stations			
Weathership E (N.Atlantic)	5.96	+2.99	0.738
Weathership V (N.Pacific)	5.51	−1.10	0.737
St.Helene (S.Atlantic)	2.80	+6.61	0.158
Diego Garcia Island (Indian Ocean)	6.93	+4.66	0.880
Midway Island (N. Pacific)	6.80	+6.15	0.840
Truk Island(N. Pacific)	7.07	+5.05	0.940

The imprint of local conditions can be seen also at a number of coastal and continental stations. The examples in Table 21 demonstrate that varying influences of different sources of vapor with different isotope characteristics, different air mass trajectories, or evaporation and isotope exchange processes below the cloud base may often lead to much more complex relationships at the local level between δD and $\delta^{18}O$ than suggested for the regional or continental scale by the global "Meteoric Water Line" equation.

Knowledge about the isotopic variations in precipitation is increased when single rain events are analyzed from local stations. Especially under mid-latitude weather conditions, such short-term variations arise from varying contributions of tropical, polar, marine, and continental air masses. These isotope data – in conjunction with other weather data – are able to provide important climatic information.

3.6.1.2
Ancient Meteoric Waters

Assuming that the H- and O-isotope compositions and temperatures of ancient ocean waters are comparable to present-day values, the isotopic composition of ancient meteoric waters may have been governed by a relation similar to the present "Meteoric Water Line." However, the application of this relationship back through time should be treated with caution. For instance, departures might be expected if humidity conditions were very different from the present situation. To date, however, there is no compelling evidence to indicate that the systematics of ancient meteoric waters were very different from the present meteoric water relationship (Sheppard 1986).

3.6.2
Ice Cores

The isotopic composition of snow and ice deposited in the polar regions and at high elevations in mountains depends primarily on the temperature. Snow deposited during the summer has less negative $\delta^{18}O$- and δD-values than snow deposited during the winter. A good example of the seasonal dependence has been given by Deutsch et al. (1966) on an Austrian glacier, where the mean δD-difference between winter and summer snow was observed to be $-14‰$. This seasonal cycle has been used to determine the annual stratigraphy of glaciers and to provide short-term climatic records. However, alteration of the snow and ice by seasonal meltwater can result in changes in the isotopic composition of the ice, thus biasing the historical climate record. Systematic isotope studies also have been used to study the flow patterns of glaciers. Profiles through a glacier should exhibit lower isotope ratios at depth than nearer the surface, because deep ice may have originated from locations upstream of the ice-core site, where temperatures should be colder.

In the past few decades, several ice cores to at least 1000 m depth have been recovered from Greenland and Antarctica. In these cores, seasonal variations are

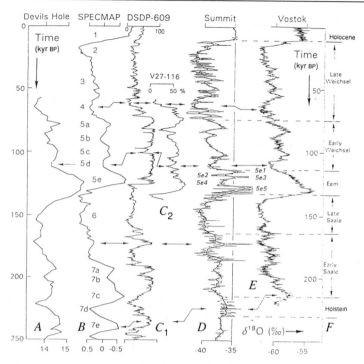

Fig. 44. Various climate $\delta^{18}O$ records of the last glacial cycles plotted on a linear timescale. *Records D and E* are from ice cores of Summit, Greenland (*GRIP*), and Vostok, Antarctica. (After Dansgaard et al. 1993)

generally observed only for the uppermost portions. After a certain depth, which depends on accumulation rates, seasonal variations completely disappear and isotopic changes are caused by long-term climatic variations. No matter how thin a sample one cuts from the ice core, its isotope composition will represent a mean value of several years of snow deposition.

The most recent ice cores – investigated in great detail by large groups of researchers – are the Vostok core from east Antarctica (Lorius et al. 1985; Jouzel et al. 1987) and the GRIP and GISP 2 cores from Greenland (Dansgaard et al. 1993; Grootes et al. 1993). In the Vostok core, the low accumulation rate of snow in Antarctica results in very thin annual layers, which means that climate changes of a century or less are difficult to resolve. The new Greenland ice cores GRIP and GISP 2 were drilled in regions with high snow accumulation near the center of the Greenland ice sheet. In these cores it is possible to resolve climate changes on the timescale of decades or less, even though they occurred 100 000 years ago. The GRIP and GISP 2 data indicate a dramatic difference between our present climate and the climate of the last interglacial period. Whereas the present interglacial climate seems to have been very stable over the last 10 000 years, the early and late parts of the last interglacial period (ca. 135 000 and ca. 115 000 years before present, respectively) were characterized by rapid fluctuations between temperatures, both warmer and very much colder than the present. It apparently took only a decade or two to shift between these very different climatic regimes.

Figure 44 compares $\delta^{18}O$ profiles from Antarctica and Greenland. The dramatic δ-shifts observed in Greenland cores are less pronounced in the δ-record along the Vostok core, probably because the shifts in Greenland are connected to rapid ocean/atmosphere circulation changes in the North Atlantic.

3.6.3
Groundwater

In temperate and humid climates the isotopic composition of groundwater is similar to that of the precipitation in the area of recharge (Gat 1971). This is strong evidence for direct recharge to an aquifer. The seasonal variation of all meteoric water is strongly attenuated during transit and storage in the ground. The degree of attenuation varies with depth and with surface and bedrock geological characteristics, but in general deep groundwaters show no seasonal variation in δD- and $\delta^{18}O$-values and have an isotopic composition close to amount-weighted mean annual precipitation values.

The characteristic isotope fingerprint of precipitation provides an effective means of identifying possible groundwater recharge areas and hence subsurface flow paths. According to Gat (1971), the main mechanisms which can cause variations between precipitation and recharged groundwater are:

1. Recharge from partially evaporated surface water bodies.

2. Recharge that occurred in past periods of different climate when the isotopic composition of precipitation was different from that at present.

3. Isotope fractionation processes resulting from differential water movement through the soil or the aquifer or due to kinetic or exchange reactions within geological formations.

In semiarid or arid regions, evaporation losses before and during recharge shift the isotopic composition of groundwater towards higher δ-values. Furthermore, transpiration of shallow groundwater through plant leaves may also be an important evaporation process. Detailed studies of soil moisture evaporation have shown that evaporation loss and isotopic enrichment are greatest in the upper part of the soil profile and are most pronounced in unvegetated soils (Welhan 1987). In some arid regions, groundwater may be classified as paleowaters, which were recharged under different meteorological conditions than present in a region today and which imply water ages of several thousand years. Gat and Issar (1974) have demonstrated that the isotopic composition of such paleowaters can be distinguished from more recently recharged groundwaters, which have been evaporated.

In summary, the application of stable isotopes to groundwater studies is based on the fact that the isotopic composition of water behaves conservatively in low-temperature environments where water-rock contact times are short relative to the kinetics of mineral-water isotope exchange reactions.

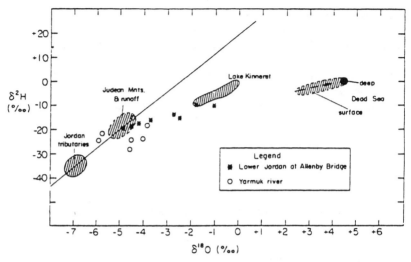

Fig. 45. δD versus δ18O of the Dead Sea and its water sources as an example of an evaporative environment. (After Gat 1984; Reprinted from Earth and Planetary Science Letters, Vol. 71, The stable isotope composition of Dead Sea waters, p 361-376, © 1984, with kind permission from Elsevier Science – NL, Sara Burgerhartstraat 25, 1055 KV Amsterdam, The Netherlands)

3.6.4
Isotope Fractionations During Evaporation

In an evaporative environment, one could expect to find extreme enrichments in the heavy isotopes D and 18O. However, this is generally not the case. Taking the Dead Sea as the typical example of an evaporative system, Fig. 45 shows only moderately enriched δ18O-values and to an even lesser degree δD-values (Gat 1984). Isotope fractionations accompanying evaporation are rather complex and can be best described by subdividing the evaporation process into several steps (Craig and Gordon 1965):

1. The presence of a saturated sublayer of water vapor at the water-atmosphere interface, which is depleted in the heavy isotopes.

2. The migration of vapor away from the boundary layer, which results in further depletion of heavy isotopes due to different diffusion rates.

3. The vapor reaching a turbulent region where mixing with vapor from other sources occurs.

4. The vapor of the turbulent zone then condensing and back reacting on the water surface.

This model qualitatively explains the deviation of isotopic compositions away from the "Meteoric Water Line" because molecular diffusion adds a nonequilibrium fractionation term and the limited isotopic enrichment occurs as a consequence of the molecular exchange with the atmospheric vapor. It is mainly the humidity which controls the degree of isotope enrichment. Only under very arid conditions, and only in small water bodies, are really large enrichments in D and

[18]O observed. For example, Gonfiantini (1986) reported a δ^{18}O-value of +31.3‰ and a δD-value of +129‰ for a small, shallow lake in the western Sahara.

3.6.5
Ocean Water

The isotopic composition of ocean water has been discussed in detail by Redfield and Friedman (1965), Craig and Gordon (1965), and Broecker (1974). Ocean water with 3.5% salinity exhibits a very narrow range in isotopic composition. There is, however, a strong correlation with salinity because evaporation which increases salinity also concentrates [18]O and D. Low salinities, which are caused by freshwater and meltwater dilution, correlate with low D and [18]O concentrations. This results in modern oceans in two nearly linear trends that meet at an inflection point where salinity is 3.55% and δ^{18}O is 0.5‰ (Fig. 46).

The high-salinity trend represents areas where evaporation exceeds precipitation, and its slope is determined by the volume and isotopic composition of the local precipitation and evaporating water vapor. However, isotope enrichments due to evaporation are limited in extent, because of back-exchange of atmospheric moisture with the evaporating fluid. Knauth and Beeunas (1986) demonstrated that the trajectory on a δD–δ^{18}O diagram taken by the residual brine depends strongly on the humidity and other climatic variables. In the case of extreme evaporation to the halite facies and beyond, experiments by Sofer and Gat (1975) indicate that progressive enrichment of the heavier isotopes does not continue indefinitely, but the trajectory hooks around.

The slope of the low salinity trend (see Fig. 46) extrapolates to a freshwater input of about –21‰ for δ^{18}O at zero salinity, reflecting the influx of high-latitude

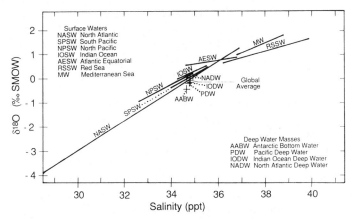

Fig. 46. Salinity versus δ^{18}O relationships in modern ocean surface and deep water masses. (After Railsback et al. 1989; Reprinted from Paleoceanography, Vol. 4 No. 5, Paleoceanographic Modeling of Temperature-Salinity Profiles From Stable Isotopic Data, pp 585-591, © 1989, with kind permission from the American Geophysical Union, 2000 Florida Avenue, NW, Washington, DC 20009, U.S.A.)

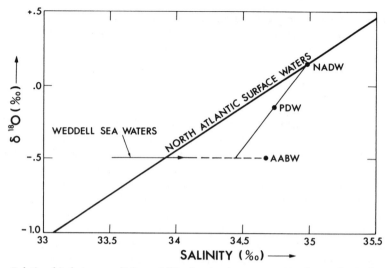

Fig. 47. Relationship between salinity and ^{18}O of major deep waters: Northern Atlantic deep water (NADW), water originating along the edge of the Antarctic continent (AABW), and Pacific deep water (PDW). PDW can only be generated by mixing roughly equal amounts of water originating in the NADW and AABW. (Broecker 1974)

precipitation and glacial meltwater. This δ-value is, in all probability, not typical of freshwater influx in nonglacial periods. Thus, the slope of the low salinity trend may have changed through geological time.

One very important aspect concerns the formation and circulation of deep water masses in the oceans. At least half of all water currently entering the deep ocean is generated in the Norwegian Sea at the northern end of the Atlantic Ocean. As isotope data indicate, North Atlantic surface waters are identical with North Atlantic deep waters. Thus, in the North Atlantic, relatively warm surface waters cool, which leads to an increase in water density and, therefore, begin to sink to abyssal depths and then flow south, across the equator, towards Antarctica into the Pacific Ocean. Joining this North Atlantic Deep Water (NADW) flow toward the Deep Pacific, is water which has been recooled in the Antarctic Ocean (AABW in Fig. 47).

Thus, deep water masses are composed of a mixture of warmer (2 °C) and saline NADW with colder (<0 °C) and less saline AABW. This relationship is shown in Fig. 47. Pacific Deep Waters (PDW) can only be generated by mixing of roughly equal amounts of water originating in the North Atlantic (NADW) and water originating along the edge of the Antarctic continent (AABW).

Another important question concerning the isotopic composition of ocean water is how constant its isotopic composition has been throughout geological history. This remains an area of ongoing controversy in stable isotope geochemistry (see Sect. 3.7). Short-term fluctuations in the isotope composition of seawater must arise during glacial periods. If all the present ice sheets in the world were melted, the δ^{18}O-value of the ocean would be lowered by about 1‰. By contrast, Fairbanks (1989) has calculated an ^{18}O-enrichment of 1.25‰ for the last maximum glaciation

3.6.6
Pore Waters

In the marine environment oxygen and hydrogen isotope compositions may be inherited from ocean water or derived from diagenetic reactions in the sediment or underlying basement. Knowledge of the chemical composition of sedimentary pore waters has increased considerably since the beginning of the Deep Sea Drilling Project. From numerous drill sites, similar depth-dependent trends in the isotopic composition have been observed.

For oxygen this means a decrease in ^{18}O from an initial δ-value very near 0‰ (ocean water) to about –2‰ at depths around 200 m (Perry et al. 1976; Lawrence and Gieskes 1981; Brumsack et al. 1992). Even lower $\delta^{18}O$-values of about –4‰ at depths of around 400 m have been observed by Matsumoto (1992). This decrease in ^{18}O is mainly due to the formation of authigenic ^{18}O-enriched clay minerals such as smectite from alteration of basaltic material and volcanic ash. Other diagenetic reactions include recrystallization of biogenic carbonates, precipitation of authigenic carbonates, and transformation of biogenic silica (opal-A) through opal-CT to quartz. The latter process, however, tends to increase $\delta^{18}O$-values. Material balance calculations by Matsumoto (1992) have indicated that the ^{18}O shift towards negative δ-values is primarily controlled by low-temperature alteration of basement basalts, which is slightly compensated by the transformation of biogenic opal to quartz.

D/H ratios may also serve as tracers of alteration reactions. Alteration of basaltic material and volcanic ash should increase δD-values of pore waters because the hydroxyl groups in clay minerals incorporate the light hydrogen isotope relative to water. However, measured δD-values of pore waters generally decrease from seawater values around 0‰ at the core tops to values that are 15‰–25‰ lower, with a good correlation between δD and $\delta^{18}O$. This strong covariation suggests that the same process is responsible for the D and ^{18}O depletion observed in many cores recovered during DSDP/ODP drilling, which is compatible with the paleo-ocean water reservoir. Quite a different process has been suggested by Lawrence and Taviani (1988) to explain the depth-dependent decrease in porewater δD-values. They proposed oxidation of local organic matter or oxidation of biogenic or mantle methane, and favored the oxidation of mantle methane, or even hydrogen, noting that oxidation of locally derived organic compounds may not be feasible because of the excessive quantity of organic material required. In conclusion, the depletion of D in porewaters is not clearly understood.

3.6.7
Formation Water

Formation waters are saline waters with salt contents ranging from ocean water to very dense Ca–Na–Cl brines. Their origin and evolution is still controversial, because the processes involved in the development of saline formation waters are complicated by the extensive changes that have taken place in the brines after sediment deposition.

Oxygen and hydrogen isotopes are a powerful tool in the study of the origin of subsurface waters. Prior to the use of isotopes, it was generally assumed that most of the formation waters in marine sedimentary rocks were of connate marine origin. This widely held view was challenged by Clayton et al. (1966), who demonstrated that waters from several sedimentary basins were predominantly of local meteoric origin.

Although formation waters show a wide range in isotopic composition, waters within a sedimentary basin are usually isotopically distinct. As is the case with surface meteoric waters, there is a general decrease in isotopic composition from low- to high-latitude settings (Fig. 48). Displacements of δD- and δ¹⁸O-values from the Meteoric Water Line (MWL) are very often correlated with salinity: the most depleted waters in D and ¹⁸O are usually the least saline, fluids most distant from the MWL tending to be the most saline.

Presently, in the view of numerous subsequent studies (i.e., Hitchon and Friedman 1969; Kharaka et al. 1974; Banner et al. 1989; Connolly et al. 1990; Stueber and Walter 1991), it is obvious that basin subsurface waters have complicated histories and frequently are mixtures of waters with different origins. The arguments of Clayton et al. (1966) were so compelling that many investigators have accepted the assumption that all connate waters in sedimentary basins are of meteoric origin.

However, as was proposed by Knauth and Beeunas (1986) and Knauth (1988), the interpretation of the isotope data from formation waters may not require complete flushing of sedimentary basins by meteoric water, but instead can result from mixing between meteoric water and the remnants of original connate waters. Knauth and Beeunas (1986) pointed out that, during the early stages of seawater

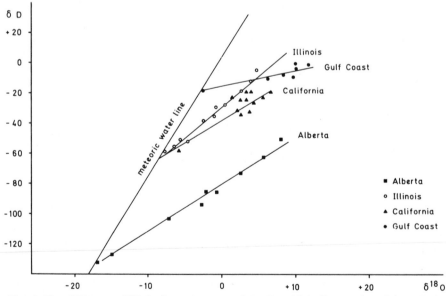

Fig. 48. Plot of δD versus δ¹⁸O for formation waters from the midcontinent region of the United States. (After Taylor 1974)

evaporation, residual water becomes enriched in ^{18}O and D, the extent depending on climatic variables (Craig and Gordon 1965 and others). Under conditions of extreme evaporation, however, the enrichment of the heavy isotopes ceases and the residual liquids become less isotopically enriched (Sofer and Gat 1975). This interpretation does not rule out other processes causing isotope fractionations.

The characteristic ^{18}O shift may be due to isotopic exchange with ^{18}O-rich sedimentary minerals, particularly carbonates. The δD-shift is less well understood, possible mechanisms for D enrichment being fractionation during membrane filtration, and exchange with H_2S, hydrocarbons, and hydrous minerals.

It is well known that shales and compacted clays can act as semipermeable membranes which prevent passage of ions in solution while allowing passage of water (ultrafiltration). Coplen and Hanshaw (1973) have shown experimentally that ultrafiltration may be accompanied by hydrogen and oxygen isotope fractionation. However, the mechanism responsible for isotopic fractionation is poorly understood. Phillips and Bentley (1987) proposed that fractionation may result from increased activity of the heavy isotopes in the membrane solution, because high cation concentrations increase hydration sphere fractionation effects. Hydrogen isotope exchange between H_2S and water will occur in nature, but probably will not be quantitatively important. Due to the large fractionation factor between H_2S and H_2O, this process might be significant on a local scale. Isotope exchange with methane or higher hydrocarbons will probably be not important, because exhange rates are extremely low at sedimentary temperatures.

Somewhat unusual isotopic compositions have been observed in highly saline deep waters from Precambrian crystalline rocks as well as in deep drill holes, which plot above or to the left of the Meteoric Water Line (Frape et al. 1984; Kelly et al. 1986; Frape and Fritz 1987). There are two major theories about the origin of these Ca-rich brines:

1. The brines represent modified Paleozoic seawater or basinal brines (Kelly et al. 1986).

2. The brines are produced by leaching of saline fluid inclusions in crystalline rocks or by intense water/rock interactions (Frape and Fritz 1987).

Possible processes that could result in isotopic compositions to the left of the Meteoric Water Line include: (1) precipitation of substantial quantities of hydrous minerals having high δ^{18}O- and low δD-values at low temperatures from limited amounts of seawater or meteoric waters, (2) the preferential exchange of hydrogen isotopes in meteoric waters having low D/H ratios with hydrous minerals having much higher D/H ratios, or (3) the loss of hydrogen from the fluids as a result of reduction of water to D-depleted methane or reduced hydrogenous gases which enrich the remaining water in deuterium.

3.6.8
Water in Hydrated Salt Minerals

Many salt minerals have water of crystallization in their crystal structure. Such water of hydration can provide information on the isotope compositions and/or temperatures of brines from which the minerals were deposited. To interpret such

Table 22. Experimentally determined fractionation factors of salt minerals and their corrections using "salt effect" coefficients. (After Horita 1989)

Mineral	Chemical formula	T (°C)	αD	$\alpha D_{(corr)}$	$\alpha^{18}O$	$\alpha^{18}O_{(corr)}$
Borax	$Na_2B_4O_7 \times 10\ H_2O$	25	1.005	1.005		
Epsomite	$MgSO_4 \times 7\ H_2O$	25	0.999	0.982		
Gaylussite	$Na_2CO_3 \times CaCO_3$					
	$\times 5\ H_2O$	25	0.987	0.966		
Gypsum	$CaSO_4 \times 2\ H_2O$	25	0.980	0.980	1.0041	1.0041
Mirabilite	$Na_2SO_4 \times 10\ H_2O$	25	1.017	1.018	1.0014	1.0014
Natron	$Na_2CO_3 \times 10\ H_2O$	10	1.017	1.012		
Trona	$Na_2CO_3 \times NaHCO_3$					
	$\times 2H_2O$	25	0.921	0.905		

isotope data, it is necessary to know the fractionation factors between the hydration water and the solution from which they are deposited. Several experimental studies have been performed to determine these fractionation factors (Gonfiantini and Fontes 1963; Matsuo et al. 1972; Matsubaya and Sakai 1973; Stewart 1974; Sofer 1978; Horita 1989). Because most saline minerals equilibrate only with highly saline solutions, the isotopic activity and isotopic concentration ratio of water in the solution are not the same (Sofer and Gat 1972). Most studies determined the isotopic concentration ratios of the mother solution and, as Horita (1989) demonstrated, these fractionation factors have to be corrected using the "salt effect" coefficients when applied to natural settings (Table 22).

3.7
Isotopic Composition of the Ocean During Geological History

The growing concern with respect to "global change" brings with it the obvious need to document and understand the geological history of sea water. From paleoecological studies it can be deduced that ocean water should not have changed its chemical composition very drastically, since marine organisms can only tolerate relatively small chemical changes in their marine environment. The similarity of the mineralogy of sedimentary rocks during the Earth's history strengthens the conclusion that the chemical composition of ocean water has not varied substantially. This relatively crude view, however, does not exclude the possibility that small changes in the chemical composition remain undetected. One of the most sensitive tracers recording the composition of ancient seawater is the isotopic composition of chemical sediments precipitated from seawater.

In the following the discussion is restricted to the stable isotopes of oxygen, carbon, and sulfur. These elements reflect different parameters such as tectonic processes (including weathering and erosion), redox conditions, stratification and circulation patterns, and past ocean temperatures. One of the most fundamental questions is which kind of sample provides the necessary information, in the sense that it represents the ocean water composition at its time of formation and has not been modified subsequently by diagenetic reactions. The quality of the measured signal also depends on the temporal resolution. The fact that sedimentary rocks have been recycled means that information on past environments is lost at a logarithmic rate.

3.7.1
Oxygen

It is generally agreed that continental glaciation and deglaciation induce small changes in the $\delta^{18}O$-value of the ocean on short time scales. There is, however, considerable debate about long-term changes. The present ocean is depleted in ^{18}O by at least 6‰ relative to the total reservoir of oxygen in the crust and mantle. Muehlenbachs and Clayton (1976) presented a model in which the isotopic composition of ocean water is held constant by two different processes: (1) low-temperature weathering of oceanic crust which depletes ocean water in ^{18}O, because ^{18}O is preferentially bound in weathering products and (2) the high-temperature hydrothermal alteration of ocean ridge basalts which enriches ocean water in ^{18}O, because ^{16}O is preferentially incorporated into the solid phase during the hydrothermal alteration of oceanic crust. If sea floor-spreading ceased, or its rate were to decline, the $\delta^{18}O$-value of the oceans would slowly change to lower values because of continued continental weathering. Gregory and Taylor (1981) presented further evidence for this rock/water buffering and showed that the $\delta^{18}O$ of seawater should be invariant within about ±1‰, assuming sea-floor spreading was operating throughout geological time. At present, it is not clear whether the sedimentary record is in accord with this model for constant oxygen isotope compositions because in a general way carbonates, cherts, and phosphates show a decrease in ^{18}O in progressively older samples (Veizer and Hoefs 1976; Knauth and Lowe 1978; Shemesh et al. 1983). The significance of these trends is still not settled, because there are three major variables that can influence the ^{18}O content of a phase that precipitates from ocean water: (1) $^{18}O/^{16}O$ ratio of the fluid, (2) temperature, and (3) diagnetic recrystallization, which can lead to obliteration of the original isotope record. Which of the three factors is dominant is a matter of ongoing debate (e.g., Land 1995 vs. Veizer 1995). While a large group of researchers considers that the ^{18}O depletion with increasing age is a result of postdepositional resetting of the primary isotopic signal, another group argues for the primary nature of the trend, with a lot of disagreement whether factor (1) or (2) is more important.

In summary, various arguments contradict each other, which leaves the issue far from being resolved.

3.7.2
Carbon

The ^{13}C content of a marine carbonate is closely related to that of the dissolved marine bicarbonate from which the carbonate is precipitated. Alteration of this primary $\delta^{13}C$-value may or may not occur during diagenetic mineral transformations, depending on the amount of carbon present in the postdepositional solutions. Changes in the ^{13}C content of bicarbonate are considered to be the result of variations in the ratio of organic carbon to carbonate carbon contributed to sediments. An increase in the amount of buried organic carbon means that ^{12}C would be preferentially removed from seawater, so that the ocean reservoir would become isotopically heavier. Thus, shifts towards higher ^{13}C contents in limestones of a given age may be due to an increase of organic carbon burial relative to carbonate carbon burial. Negative $\delta^{13}C$-shifts accordingly may indicate a decrease in the rate of carbon burial.

Because of the relationship between carbonate and organic carbon, a parallel shift in the isotope composition of both carbon reservoirs should be observed. Although such covariation in C-isotope composition has been observed, it is not a ubiquitous feature (Hoefs 1981). Very often the isotope composition of organic matter is more variable (i.e., Dean et al. 1986), which suggests that the record of C_{org} is influenced by secondary processes such as the diagenetic transformation of primary organic matter into secondary kerogen. Therefore, the discussion which follows will be restricted to carbonates.

$\delta^{13}C$-values of limestones vary mostly within a band of $0\pm3\permil$ since at least 3.5 Ga (Veizer and Hoefs 1976; Schidlowski et al. 1983). The long-term C-isotope trend for carbonates can be punctuated by sudden shifts over very short time intervals named "carbon isotope events" which are considered to represent characteristic features, and have been used as time markers in stratigraphic correlations.

Large C-isotope changes in carbonates have been measured at the Precambrian/Cambrian, Permian/Triassic, and Cretaceous/Tertiary boundaries (Magaritz 1991). The shape of the curve at these three events shows (1) a decrease in $\delta^{13}C$-values toward the boundary, (2) a minimum in the curve sometimes following the stratigraphic horizon considered to represent the time boundary, and (3) an increase in the $\delta^{13}C$-value, which can be associated with an increase in productivity in the ocean. Of all of these C-isotope curves, the terminal Permian shows the longest minimum interval prior to the increase toward a new level, suggesting a rather long period of reduced photosynthetic activity. Although the shape of the boundary curves of these three events exhibits similarities, it is not at all clear that the causes are the same (Holser and Magaritz 1992).

Very enriched $\delta^{13}C$-values up to $10\permil$ and higher have been measured for carbonates at about 2.2–2.0 Ga of age and at the end of the Proterozoic, both of which may represent periods of increased burial of organic carbon (Knoll et al. 1986; Baker and Fallick 1989; Derry et al. 1992, and others). The cycles of carbon and sulfur are linked by biologic oxidation and reduction processes. For both C and S, the major fractionations occur during biologically mediated reduction. As noticed by Veizer et al. (1980), $\delta^{13}C$-variations in carbonates tend to be inverse to those of $\delta^{34}S$ of sulfates, although the mechanism which drives this balancing remains unclear.

3.7.3
Sulfur

The best-documented trend of isotope variations through time is that for the sulfur isotope composition of marine sulfate. Because isotope fractionation between dissolved sulfate in ocean water and gypsum/anhydrite is negligible (Raab and Spiro 1991), evaporite sulfates closely reflect the sulfur isotope composition of marine sulfate through time. The first S-isotope "age curves" were published by Nielsen and Ricke (1964) and Thode and Monster (1964). Since then, this curve has been updated by many more analyses (Holser and Kaplan 1966; Nielsen 1972; Holser 1977; Claypool et al. 1980) (see Fig. 49). The sulfur isotope curve varies from a maximum of $\delta^{34}S = +30‰$ in early Paleozoic time, to a minimum of $+10‰$ in Permian time, to near $+16‰$ during most of the Mesozoic. These shifts are considered to reflect net fluxes of isotopically light sulfur during bacterial reduction of oceanic sulfate to the reservoir of reduced sulfide in sediments, thus increasing the ^{34}S content in the remaining oceanic sulfate reservoir. Conversely, a net return flux of the light sulfide to the ocean during weathering decreases marine sulfate $\delta^{34}S$-values. From this relationship, it might be expected that a parallel age curve to that for sulfates should exist for sedimentary sulfides. However, the available S-isotope data for sulfides range widely and seem strongly dependent on the degree to which the reduction system is "open" and on the sedimentation rate so that age trends are obscured.

Accepting that a difference in $\delta^{34}S$-values of 40–60‰ between bacteriogenic sulfide and marine sulfate exists in present-day sedimentary environments, simi-

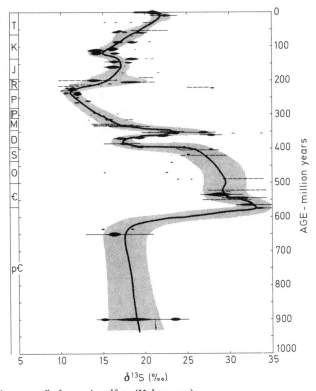

Fig. 49. $\delta^{34}S$ "age curve" of oceanic sulfate. (Holser 1977)

lar fractionations in ancient sedimentary rocks may be interpreted as evidence for the activity of sulfate-reducing bacteria. The presence or absence of such fractionations in sedimentary rocks thus may constrain the time of emergence of sulfate-reducing bacteria. In early Archean sedimentary rocks, most sulfides and the rare sulfates have $\delta^{34}S$-values near 0‰ (Monster et al. 1979; Cameron 1982), which has been interpreted as indicating an absence of bacterial reduction in the Archean. Recently, Ohmoto et al. (1993) employed a laser microprobe approach to analyze single pyrite grains from the Barberton formation and observed a variation of up to 10‰ among pyrites from a single small rock specimen, which could imply that bacterial reduction occurs since at least 3.4 Ga. Whatever the actual causes may be for the fluctuations of the $\delta^{34}S$-values during the geological past, it is obvious that they are not compatible with the concept of a steady-state ocean, where the partitioning into reduced and oxidized reservoirs would be at a fixed ratio.

In conclusion, the observed variations in the isotope composition of carbon and sulfur argue against a constant chemical composition of the ocean during the Earth's history and favor sizable temporal variations.

3.8
Isotopic Composition of Dissolved and Particulate Compounds in Ocean and Fresh Waters

The following section will discuss the carbon, nitrogen, oxygen, and sulfur isotope composition of dissolved and particulate compounds in ocean and fresh waters. The isotopic compositions of dissolved components in waters of different origins depends on a variety of processes such as the composition of the minerals which have been dissolved during weathering, the inorganic or organic nature of the precipitation process, and exchange with the gases of the atmosphere. Of special importance are biological processes acting mainly in surface waters, which tend to deplete certain elements such as carbon and nitrogen in surface waters by biological uptake, and which subsequently are returned at depth by oxidation processes.

Most particles in seawater come from organisms, with large particles sinking rapidly and being less degraded by chemical and biological processes than small particles as they sink through the water column.

3.8.1
Carbon Species in Water

3.8.1.1
Bicarbonate in Ocean Water

In addition to organic carbon, four other carbon species exist in natural water: dissolved CO_2, H_2CO_3, HCO_3^-, and CO_3^{2-}, all of which tend to equilibrate as a function of temperature and pH. HCO_3^- is the dominant C-bearing species in ocean water. A typical ocean water C-isotope vertical profile is shown in Fig. 50. Most surface waters in the central ocean basins have $\delta^{13}C$-values of about 2.2‰ (Deuser and Hunt 1969; Kroopnick 1985). However, this value decreases into deeper

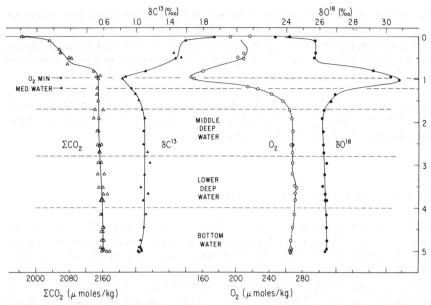

Fig. 50. Vertical profiles of dissolved ΣCO_2, $\delta^{13}C$, dissolved O_2, and $\delta^{18}O$ in the North Atlantic. (Kroopnick et al. 1977)

water masses due to continuous flux of ^{13}C-depleted organic and skeletal detritus, and its subsequent oxidation, as it falls through the water column.

North Atlantic Deep Water (NADW), which is formed with an initial $\delta^{13}C$-value between 1.0 and 1.5‰, becomes gradually lower in ^{13}C as it travels southward and mixes with Southern Ocean Water, which has an average $\delta^{13}C$-value of 0.3‰ (Kroopnick 1985). As this deep water travels to the Pacific Ocean, its $^{13}C/^{12}C$ ratio is further reduced by 0.5‰ by the continuous oxidation of organic matter. This is the basis for using $\delta^{13}C$-values as a tracer of paleo-oceanographic changes in deep water circulation (i.e., Curry et al. 1988). Another factor which influences the $\delta^{13}C$-value of the ocean is the addition of anthropogenic CO_2. Quay et al. (1992) have demonstrated that the $\delta^{13}C$-value of dissolved bicarbonate in the surface waters of the Pacific has decreased by about 0.4‰ between 1970 and 1990. If this number is valid for the ocean as a whole, it would allow a quantitative estimate for the net oceanic CO_2 uptake and a quantification for the net sink of anthropogenically produced CO_2.

3.8.1.2
Particulate Organic Matter

Particulate organic matter (POM) in the ocean originates in large part from the detrital remains of plankton in the euphotic zone and reflects living plankton populations. As POM sinks, biological reworking changes its chemical composition, the extent of this reworking depending on residence time in the water column. Most POM profiles described in the literature exhibit a general trend of surface isotopic values comparable to those for living plankton, with δ^{13}C-values becoming increasingly lower with depth. Eadie and Jeffrey (1973) and Jeffrey et al. (1983) interpreted this trend as the loss of labile, ^{13}C-enriched amino acids and sugars through biological reworking which leaves behind the more refractory, isotopically light lipid component.

C/N ratios of POM increase with depth of the water column. This implies that nitrogen is more rapidly lost than carbon during degradation of POM. This is the reason for the much greater variation in δ^{15}N-values than in δ^{13}C-values (Saino and Hattori 1980; Altabet and McCarthy 1985).

3.8.1.3
Carbon Isotope Composition of Pore Waters

Initially the pore water at the sediment/water interface has a δ^{13}C-value near that of seawater. In sediments, the decomposition of organic matter consumes oxygen and releases isotopically light CO_2 to the pore water, while the dissolution of Ca-CO_3 adds CO_2 which is isotopically heavy. The carbon isotope composition of pore waters at a given locality and depth should reflect modification by the interplay of these two processes. The net result is to make pore waters isotopically lighter than the overlying bottom water (Nissenbaum et al. 1972; Grossman 1984). Mc-Corkle et al. (1985) and McCorkle and Emerson (1988) have shown that steep gradients in pore water δ^{13}C-values exist in the first few centimeters below the sediment-water interface. The observed δ^{13}C-profiles vary systematically with the rain of organic matter to the sea floor, with higher carbon rain rates resulting in isotopically lighter δ^{13}C-values (Fig. 51).

One would expect that pore waters would have ^{13}C/^{12}C ratios no lower than organic matter. However, a more complex situation is actually observed due to bac-

Fig. 51. δ^{13}C records of total dissolved CO_2 from pore waters of anoxic sediments recovered in various Deep Sea Drilling Sites. (Anderson and Arthur 1983)

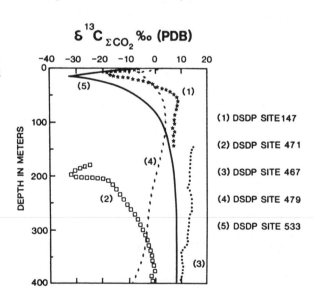

(1) DSDP SITE 147

(2) DSDP SITE 471

(3) DSDP SITE 467

(4) DSDP SITE 479

(5) DSDP SITE 533

terial methanogenesis. Bacterial methane production generally follows sulfate reduction in anaerobic carbon-rich sediments, the two microbiological environments being distinct from one another. Since methane-producing bacteria produce very ^{12}C-rich methane, the residual pore water can become significantly enriched in ^{13}C. As bacterial methane production continues, the pore waters evolve to higher $\delta^{13}C$-values (see Fig. 51). However, the trends shown in Fig. 51 cannot solely be interpreted in terms of amounts of sulfate reduction and methane formation; additional factors such as carbon losses from and gains to the pore water system also have to be taken into account.

3.8.1.4
Carbon in Fresh Waters

Dissolved carbonate in fresh waters may exhibit an extremely variable isotopic composition, because it represents varying mixtures of carbonate species derived from weathering of carbonates and that originating from biogenic sources such as freshwater plankton or CO_2 from bacterial oxidation of organic matter in the water column or in soils (Hitchon and Krouse 1972; Longinelli and Edmond 1983; Pawellek and Veizer 1994; Cameron et al. 1995). The CO_2 partial pressures in rivers vary widely around the equilibrium value with the atmosphere. For instance, the Danube shows considerable seasonal variations in CO_2: in winter dissolved inorganic carbon (DIC) partially equilibrates with atmospheric CO_2, in summer a progressive ^{13}C depletion of DIC points to bacterial respiration of biogenic carbon (Pawellek and Veizer 1994). Comparison of the data from the MacKenzie River (Hitchon and Krouse 1972) with those from the Amazon Basin (Longinelli and Edmond 1983) reveals an interesting difference. The MacKenzie River data have a $\delta^{13}C$-peak at about −9‰, whereas the Amazon River data are displaced to about −20‰ with a broad distribution range. These differences are consistent with a dominance of carbonate weathering in the MacKenzie River drainage system, whereas in the tropical environment of the Amazon River biological CO_2 predominates.

3.8.2
Nitrogen

Nitrogen is one of the limiting nutrients in the ocean. Apparently, the rate of nitrate formation is so slow, and the denitrification in the ocean so rapid, that nitrogen is in short supply. Dissolved nitrogen is subject to isotope fractionation during microbial processes and during biological uptake. Nitrate dissolved in deep water has a $\delta^{15}N$-value of 6–8‰ (Cline and Kaplan 1975; Wada and Hattori 1976). Denitrification seems to be the principal mechanism that keeps marine nitrogen at higher $\delta^{15}N$-values than atmospheric nitrogen.

The $\delta^{15}N$-value of particulate material was originally thought to be determined by the relative quantities of marine and terrestrial organic matter. However, temporal variations in the ^{15}N content of particulate matter predominate and obscure N-isotopic differences previously used to distinguish terrestrial from marine organic matter. Altabet and Deuser (1985) observed seasonal variations in particles sinking to the ocean bottom and suggested that $\delta^{15}N$-values of sinking particles represent a monitor for nitrate flux in the euphotic zone. Natural ^{15}N-variations can thus provide information for the vertical structure of nitrogen cycling in the ocean.

Saino and Hattori (1980) first observed distinct vertical changes in the ^{15}N content of suspended particulate nitrogen and related these changes to particle diagenesis. A sharp increase in ^{15}N below the base of the euphotic zone has been ubiquitously observed (Altabet and McCarthy 1985; Saino and Hattori 1987; Altabet 1988). These findings imply that the vertical transport of organic matter is mediated primarily by rapidly sinking particles and that most of the decomposition of organic matter takes place in the shallow layer beneath the bottom of the euphotic zone.

3.8.3
Oxygen

As early as 1951, Rakestraw et al. demonstrated that dissolved oxygen in the oceans is enriched in ^{18}O relative to atmospheric oxygen. Extreme enrichments up to 14‰ (Kroopnick and Craig 1976) occur in the oxygen minimum region of the deep ocean due to preferential consumption of ^{16}O by bacteria in abyssal ocean waters, which is evidence for a deep metabolism (see Fig. 50).

3.8.4
Sulfate

Modern ocean water has a fairly constant $\delta^{34}S$-value of 21‰ (Rees et al. 1978) and $\delta^{18}O$-value of 9.6‰ (Lloyd 1967, 1968; Longinelli and Craig 1967). From theoretical calculations of Urey (1947), it is quite clear that the $\delta^{18}O$-value of dissolved sulfate does not represent equilibrium with $\delta^{18}O$-value of the water, but how this value has been achieved is still controversial. Lloyd (1967, 1968) proposed a model in which the fast bacterial turnover of sulfate at the sea bottom determines the oxygen isotope composition of dissolved sulfate. This conclusion was questioned by Holser et al. (1979), who argued that the oxygen isotope composition of seawater sulfate should be controlled by a dynamic balance of sulfate inputs (mainly from weathering of sulfides and sulfates) and sulfate outputs (mainly through evaporite formation and sulfate reduction).

In freshwater environments, the sulfur and oxygen isotope composition of dissolved sulfate is much more variable and potentially the isotope ratios can be used to identify the sources. However, such attempts have been only partially successful because of the variable composition of the different sources. $\delta^{34}S$-values of different rivers and lakes show a rather large spread as is demonstrated in Fig. 52. The data of Hitchon and Krouse (1972) for water samples from the MacKenzie River drainage system exhibit a wide range of $\delta^{34}S$-values, reflecting contributions from marine evaporites and shales. Surprisingly, Longinelli and Edmond (1983) found a very narrow range for the Amazon River, which was interpreted as representing a dominant Andean source of Permian evaporites with a lesser admixture of sulfide sulfur. Rabinovich and Grinenko (1979) reported time-series measurements for the large European and Asian rivers in Russia. The sulfur in the

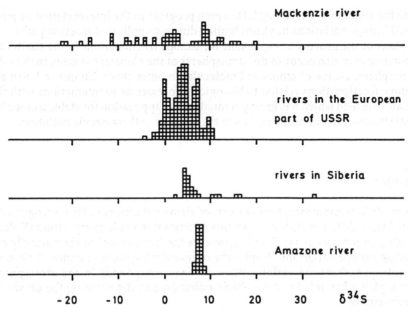

Fig. 52. Frequency distribution of δ^{34}S-values in river sulfate

European river systems should be dominated by anthropogenically derived sources, which in general have δ^{34}S-values between 2 and 6‰.

The oxygen isotope composition of freshwater sulfate can be highly variable too. Cortecci and Longinelli (1970) and Longinelli and Bartelloni (1978) observed a δ^{18}O range from 5‰ to 19‰ in rainwater samples from Italy and postulated that most of the sulfate is not oceanic in origin, but rather produced by oxidation of sulfur during the burning of fossil fuels. The oxidation of reduced sulfur to sulfate is a complex process which involves chemical and microbiological aspects. Two general pathways of oxidation have been suggested: (1) oxidation by molecular oxygen and (2) oxidation by ferric iron plus surface water. The large difference in oxygen isotope composition between atmospheric oxygen and surface water should permit a distinction between these two processes. In reality, intermediate steps common to both pathways might exist. Schwarcz and Cortecci (1974) suggested that sulfate derived from oxidation of reduced sulfur contains approximately equal amounts of water and molecular oxygen.

3.9
Atmosphere

In recent years tremendous progress has been achieved in the analysis of the isotope composition of important trace compounds in the atmosphere. The major elements – nitrogen, oxygen, carbon – continually break apart and recombine in a multitude of photochemical reactions, which have the potential to produce isotope fractionations (Kaye 1987). However, progress in the interpretation of measured isotope variations has been limited due to complicated reaction paths.

Many of the processes responsible for isotope fractionations in the Earth's atmosphere may also occur in the atmospheres of the planetary system, such as the atmospheric escape of atoms and molecules to outer space. Unique to Earth are isotope fractionations related to biological processes or to interactions with the ocean. One field which has great potential for the application for stable isotope investigations of the atmosphere lies in the study of anthropogenic pollution.

3.9.1
Nitrogen

Nearly 80% of the atmosphere consists of elemental nitrogen. This nitrogen, collected from different altitudes, exhibits a constant isotopic composition (Dole et al. 1954; Sweeney et al. 1978) and represents the "zero-point" of the naturally occurring isotope variations. Besides the overwhelming predominance of elemental nitrogen, there are various other nitrogen compounds in the atmosphere, which play a key role in atmospheric pollution and determining the acidity of precipitation.

Nitrate originates from gaseous emissions of NO_x ($NO+NO_2$). Heaton (1986) has discussed the possibility of isotopically differentiating between naturally produced and anthropogenic NO_x. Since very little isotope fractionation is envisaged at the high temperatures of combustion in power plants and vehicles, the $\delta^{15}N$-value of pollution nitrate is expected to be similar to that of the nitrogen which is oxidized. In soils, NO_x is produced by nitrification and denitrification processes which are kinetically controlled. This, in principle, should lead to more negative $\delta^{15}N$-values in natural nitrate compared to anthropogenic nitrate. However, Heaton (1986) concluded that this distinction cannot be made on the basis of ^{15}N-contents, which has been confirmed by Durka et al. (1994). The latter authors demonstrated, however, that the oxygen isotope composition of nitrate is more indicative. Industrially produced nitrate contains oxygen from the atmosphere ($\delta^{18}O$-values of 23.5‰), while nitrate originating from a nitrification process must have water as the main oxygen source (Amberger and Schmidt 1987).

Besides NO_x oxides, there is nitrous oxide (N_2O), which is of special interest in isotope geochemistry. N_2O is an important greenhouse gas that is – on a molecular basis – a much more effective contributor to global warming than CO_2. Nitrous oxide is produced primarily by bacterial processes and is destroyed photochemically in the stratosphere. The first $\delta^{15}N$-values for N_2O were determined by Yoshida et al. (1984), and the first $\delta^{18}O$-values were published by Kim and Craig (1990). Isotope data from different soil sites presented by Kim and Craig (1993) show rather large variations with $\delta^{15}N$-values lower than for tropospheric air and oceanic samples. Thus, there must exist sources of N_2O that are enriched in ^{15}N to balance the input fluxes of lighter soil-gas N_2O. The processes that determine the ultimate nitrogen and oxygen isotope ratios have not been identified unambiguously. One potentially significant anthropogenic source of N_2O is as a by-product in the manufacture of nylon (Thiemens and Trogler 1991).

3.9.2
Oxygen

Atmospheric oxygen has a rather constant isotopic composition with a $\delta^{18}O$-value of 23.5‰ (Dole et al. 1954; Kroopnick and Craig 1972; Horibe et al. 1973). Urey (1947) calculated that if equilibrium was obtained between atmospheric oxygen and water, then atmospheric oxygen should be enriched in ^{18}O by 6‰ at 25 °C. This means atmospheric oxygen cannot be in equilibrium with the hydrosphere and thus the ^{18}O enrichment of atmospheric oxygen, the so-called "Dole" effect, must have another explanation.

It is generally agreed that this ^{18}O enrichment is of biological origin, in that photosynthesis produces ^{18}O-enriched oxygen while respiration consumes ^{18}O (Lane and Dole 1956). Kroopnick (1975) measured the oxygen isotope fractionation during respiration of natural plant populations in ocean water and found that respiration can produce an ^{18}O enrichment of about 21‰. It is therefore reasonable to assume that the $\delta^{18}O$-value of atmospheric oxygen is balanced between input from photosynthesis and output by respiration.

Sowers et al. (1991) have pioneered the analysis of $\delta^{18}O$ of O_2 in air bubbles trapped in ice cores. They examined the response of the terrestrial and marine biomass to climate change by measuring the difference between the $\delta^{18}O$-values of atmospheric oxygen and ocean water, and documented that the variability of the Dole effect is small between glacial and interglacial periods. Observed variations in the ^{18}O contents during the past 130 000 years follow the $\delta^{18}O$-value of seawater because photosynthesis transmits variations in ^{18}O of seawater to O_2 in air.

3.9.3
Ozone

Ozone has become one of the most important molecules in atmospheric research. In situ mass-spectrometric measurements by Mauersberger (1981, 1987) demonstrated that an enrichment in ^{17}O and ^{18}O of about 40% exists in the stratosphere, with a maximum at about 32 km. The rate of formation of partially isotopically substituted ozone (mass 50) is obviously faster than that of unsubstituted ozone (mass 48). This mass-independent fractionation has also been observed in laboratory experiments by Thiemens and Heidenreich (1983); however, many of the stratospheric ozone enrichments are significantly larger than those observed in laboratory experiments (Thiemens et al. 1995).

3.9.4
Carbon Dioxide

3.9.4.1
Carbon

The increasing CO_2 content of the atmosphere is a problem of world-wide concern. By measuring both the concentration and isotope composition of CO_2 on the same samples of air, it is possible to determine whether variations are of anthropogenic, oceanic, or biologic origin. The first extensive measurements of the carbon isotope ratio of CO_2 were made in 1955/1956 by Keeling (1958, 1961). He noted daily, seasonal, secular, local, and regional variations as regular fluctuations. Daily variations exist over continents, which depend on plant respiration and reach a distinct maximum around midnight or in the early morning hours. At night there is a measurable contribution of respiratory CO_2, which shifts $\delta^{13}C$-values toward lighter values (see Fig. 53). Seasonal variations in ^{13}C are very similar to CO_2 concentrations and result from terrestrial plant activity. As shown in Fig. 54 the seasonal cycle diminishes from north to south, as expected from the greater seasonality of plant activity at high latitude. Due to the greater amount of land area in the northern hemisphere, this effect is hardly discernible in the southern hemisphere (Keeling et al. 1989).

Long-term measurements of atmospheric CO_2 have been available for a few clean-air locations on an almost continuous basis since 1978 (Keeling et al. 1979, 1984, 1989, 1995; Mook et al. 1983; Ciais et al. 1995). These measurements clearly demonstrate that on average atmospheric CO_2 increases by about 1.5 ppm/year while the isotope ratio shifts toward lower $^{13}C/^{12}C$ ratios. The annual combustion of 10^{15} g fossil fuel with an average $\delta^{13}C$-value of -27% would change the ^{13}C content of atmospheric CO_2 by -0.02%/year. The observed change is, however, much smaller. Of the CO_2 emitted into the atmosphere, roughly half remains in the atmosphere and the other half is absorbed into the oceans and the terrestrial biosphere. The partitioning between these two sinks is a matter of debate. Whereas most oceanographers argue that the oceanic sink is not large enough to account for the entire absorption, terrestrial ecologists doubt that the terrestrial biosphere can be a large carbon sink. Carbon isotope measurements can be used to estimate the relative contributions of terrestrial and oceanic source and sink processes.

By comparing the average carbon isotope composition of recent foraminifera with older foraminifera, Beveridge and Shackleton (1994) have postulated that the ^{13}C content of dissolved bicarbonate in surface waters has decreased in response to

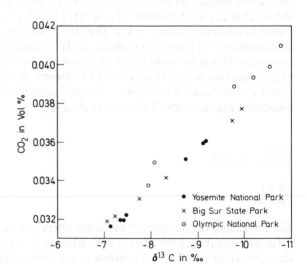

Fig. 53. Relationship between atmospheric CO_2 concentration and $\delta^{13}C_{CO_2}$.
(After Keeling 1958)

Fig. 54. Seasonal $\delta^{13}C$ variations of atmospheric CO_2 from five northern hemisphere stations. *Dots* denote monthly averages, *oscillating curves* are fits of daily averages. (After Keeling et al. 1989; Reprinted from Geophysical Monographs Book Series, Vol. 55 A three dimensional model of atmospheric CO2 transport based on observed winds, pp 165-236, © 1989, with kind permission from the American Geophysical Union, 2000 Florida Avenue, NW, Washington, DC 20009, U.S.A.)

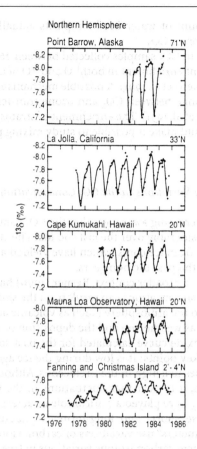

an invasion of anthropogenic CO_2. Estimates of the magnitude of decrease seem to depend on the depth at which the foraminifera species live. Ciais et al. (1995) studied weekly CO_2 samples from a global network of 43 sites which defined the latitudinal and temporal patterns of the two carbon sinks. They observed a strong terrestrial biospheric sink in the temperate latitudes of the Northern Hemisphere.

3.9.4.2
Oxygen

Atmospheric CO_2 has a $\delta^{18}O$-value of about +41‰, which means that atmospheric CO_2 is in approximate isotope equilibrium with ocean water, but not with atmospheric oxygen (Keeling 1961; Bottinga and Craig 1969). Later measurements by Mook et al. (1983) and Francey and Tans (1987) have revealed large-scale seasonal and regional variations. Particularly noteworthy is the observation that the ^{18}O content of CO_2 is lower at high latitudes in the Northern Hemisphere. Near 70°N where precipitation is very depleted in ^{18}O, photosynthesis and respiration both lead to ^{18}O depletions in CO_2. Recently, Farquhar et al. (1993) argued that the very small amount of water in chloroplasts actually dominates the oxygen isotope composition of CO_2.

Oxygen in CO_2 samples collected between 26 and 35 km altitude show a mass-independent enrichment in both[17]O and [18]O of about 11‰ above tropospheric values (Thiemens et al. 1991). A possible mechanism for this enrichment might be isotope exchange between CO_2 and ozone transferring the enrichment of ozone to CO_2 (Yung et al. 1991). The enrichment of stratospheric CO_2 relative to tropospheric CO_2 should make it possible to study mixing processes across the tropopause.

3.9.4.3
Long-Term Variations in the CO_2 Concentration

There is increasing awareness that the CO_2 content of the Earth's atmosphere has varied considerably over the last 500 Ma. The clearest evidence comes from measurements from ice cores, which have yielded an impressive record of CO_2 variations over the past 160 000 years.

In a much broader context, Berner (1990) has modeled how long-term changes in CO_2 concentrations can result from the shifting balance of processes that deliver CO_2 to the atmosphere (such as volcanic activity) and processes that extract CO_2 (such as weathering and the deposition of organic material). The theoretical carbon dioxide curve calculated for the past 500 Ma matches the climate record at several key points: it is low during the ice age of the Carboniferous and Permian and rises to a high in the Cretaceous. Although the exact curve is far from being known, it is clear that fluctuations in the CO_2 content of the ancient atmosphere may have played a critical role in determining paleotemperatures. To elucidate these short- and long-term CO_2 fluctuations, several promising "CO_2-paleobarometers" use variations of carbon isotopes in different materials.

Short-term carbon isotope variations in tree rings have been interpreted as indicators of anthropogenic CO_2 combustion (Freyer 1979; Freyer and Belacy 1983). While different trees show a wide variability in their isotope records due to climatic and physiological factors, many tree-ring records indicate a 1.5‰ decrease in $\delta^{13}C$-values from 1750 to 1980. Freyer and Belacy (1983) reported C-isotope data for the past 500 years on two sets of European oak trees: forest trees exhibit large nonsystematic ^{13}C variations over the 500 years, whereas free-standing trees show smaller ^{13}C fluctuations which can be correlated to climatic changes. Since industrialization of these areas in 1850, the ^{13}C record for the free-standing trees has been dominated by a systematic decrease of about 2‰.

The most convincing evidence for changes in atmospheric CO_2 concentrations and $\delta^{13}C$-values comes from air trapped in ice cores in Antarctica. During the last ice age, with low CO_2 concentrations atmospheric CO_2 was isotopically lighter by about 0.3‰ relative to interglacial periods (Leuenberger et al. 1992). This somewhat surprising feature, which is opposite to the recent anthropogenic trend, is explained by either a decrease in dissolved CO_2 in surface waters because of a more efficient "biological pump" or a higher alkalinity in the glacial ocean.

Two different classes of approaches have been used in the study of long-term atmospheric CO_2 change: one utilizing deep-sea sediments, the other studying continental sediments. Cerling (1991) has been reconstructing the CO_2 content of the ancient atmosphere by analyzing fossil soil carbonate that formed from CO_2 diffusion from the atmosphere or plant roots. This method relies on certain assumptions and prerequisites. One, for instance, is the necessity of differentiating pedogenic calcretes from groundwater ones, which cannot be used for pCO_2 determinations. Yapp and Poths (1992) have analyzed sedimentary goethite, which incorporates small quantities of CO_2 in its structure. A very promising approach uses the observed relationship between the concentration of molecular CO_2 and the $\delta^{13}C$-value of organic plankton (Rau et al. 1992). This approach requires the measurement of dissolved CO_2 in the form of planktonic foraminifera and measurement of selected organic molecules, for instance alkenones (Jasper and Hayes 1990). It complements earlier attempts of Shackleton et al. (1983), who have obtained a record on pCO_2 from the $\delta^{13}C$ difference between surface and deep water foraminifera.

3.9.5
Methane

Methane enters the atmosphere from biological and anthropogenic sources and is destroyed by reaction with the hydroxyl radical. Thus, a mass-weighted average composition of all CH_4 sources is equal to the mean $\delta^{13}C$-value of atmospheric methane, corrected for any isotope fractionation effects in CH_4 sink reactions. Based on the concentration measured in air contained in polar ice cores, methane concentrations have doubled over the past several hundred years (Stevens 1988).

Methane is produced by bacteria under anaerobic conditions in wet environments such as wetlands, swamps, and rice fields. It is also produced in the stomachs of cattle and possibly by termites. Typical anthropogenic sources are fossil fuels such as coal mining and as a by-product in the burning of biomass.

Atmospheric methane has a mean $\delta^{13}C$-value of $-47‰$ (Stevens 1988). There are seasonal variations and a systematic difference between both hemispheres (Quay et al. 1991). Methane extracted from air bubbles in polar ice up to 350 years in age has a $\delta^{13}C$-value which is 2‰ lower than at present (Craig et al. 1988). This may indicate that anthropogenic burning of the Earth's biomass may be the principal cause of the recent ^{13}C enrichment in methane.

3.9.6
Hydrogen

The hydrogen isotope geochemistry of the atmosphere is very complex, because there are numerous hydrogen-containing compounds undergoing continuous chemical and physical transformations. Many studies of the isotope composition of H_2 have been performed in conjunction with the measurement of atmospheric tritium. The major result from these studies is the large variability in deuterium with both time and location, the best estimate being in the vicinity of 70‰±30‰ (Friedman and Scholz 1974). This high δD-value can be ascribed to the presence of two hydrogen components: a "background" component with enhanced deuterium and tritium and a locally produced "industrial" component which is very depleted in D.

3.9.7
Sulfur

Sulfur is found in trace compounds in the atmosphere, where it occurs in aerosols as sulfate and in the gaseous state as H_2S, and SO_2. Sulfur can orginate naturally (volcanic, sea spray, aeolian weathering, biogenic) or anthropogenically (combustion and refining of fossil fuels, ore smelting, gypsum processing). These different sources differ greatly in their isotopic composition as shown in Fig. 55. The complexities involved in the isotopic composition of atmospheric sulfur have been discussed in a recent book (*Scope 43*, edited by Krouse and Grinenko 1991).

Summarizing, the isotopic composition of the industrial sulfur sources are generally so variable that the assessment of anthropogenic contributions to the atmosphere is extremely difficult. Krouse and Case (1983) were able to give semi-quantitative estimates for a very unique situation in Alberta where the industrial SO_2 had a constant $\delta^{34}S$-value near 20‰. Generally, the premises are much more complicated, which limits the "fingerprint" character of the sulfur isotope composition of atmospheric sulfur to such rare cases.

Very interesting seasonal dependencies for sulfur in precipitation and in aerosol samples have been observed by Nriagu and Coker (1978) and Nriagu et al. (1991). $\delta^{34}S$-data for aerosol samples of the Canadian Arctic show pronounced seasonal differences, with the sulfur being more ^{34}S enriched in summer than in winter. This situation is quite different from that observed for airborne sulfur in

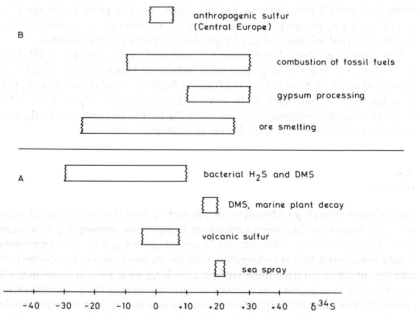

Fig. 55A,B. S-isotope composition of natural (A) and anthropogenic (B) sources of atmospheric sulfur compounds. *DMS*, dimethylsulfide

southern Canada. In rural and remote areas of southern Canada, the δ^{34}S-values of atmospheric samples are higher in winter and lower in summer. While sulfur during the winter is mainly derived from space heating and industrial sources, in summer the large emission of ^{34}S-depleted biogenic sulfur from soils, vegetation, marshes, and wetlands results in the lowering of the δ^{34}S-values of airborne sulfur. The opposite trend observed for aerosol sulfur in the Arctic suggests a different origin of the sulfur in that high-latitude area.

3.10
Biosphere

As used here, the term "biosphere" includes the total sum of living matter – plants, animals, and microorganisms and the residues of the living matter in the geological environment such as coal and petroleum. A fairly close balance exists between photosynthesis and respiration, although over the whole of geological time respiration has been exceeded by photosynthesis, and the energy derived from this was stored mostly in disseminated organic matter, and, of course, in coal and petroleum. Questions concerning the origin of coal and petroleum center on three topics: the nature and composition of the parent organisms, the mode of accumulation of the organic material, and the reactions whereby this material was transformed into the end products.

Petroleum (frequently also called crude oil) is a naturally occurring complex mixture, composed mainly of hydrocarbons, but also containing varying amounts of heterocompounds containing S, N, O, and metalloorganic molecules, such as vanadium- and nickel-porphyins. Although there are, without any doubt, numerous compounds that have been formed directly from biologically produced molecules, the majority of petroleum components are of secondary origin, either decomposition products or products of condensation and polymerization reactions.

3.10.1
Living Organic Matter

3.10.1.1
Bulk Carbon

The complexities involved in the photosynthetic fixation of carbon have already been discussed briefly on p. 41. Wickman (1952) and Craig (1953) were the first to demonstrate that marine plants are about 10‰ enriched in ^{13}C relative to terrestrial plants. Since that time numerous studies have broadened this view and provided a much more detailed picture of isotope variations in the biosphere. The large C-isotope differences found in plants were only satisfactorily explained after the discovery of new photosynthetic pathways in the 1960s. The bulk of the plant kingdom fixes CO_2 during the pathway described by Calvin (also called C_3 pathway). The two additional pathways are known as Hatch-Slack (or C_4 pathway) and CAM (Crassulacean acid metabolism – diurnal process of acidification and deacidification). The differences in isotopic composition characteristic for each one of the pathways are due to different enzymatic processes and different sizes of the metabolic pools of carbon.

Figure 56 summarizes the variability of $\delta^{13}C$-values exhibited by some major groups of higher plants, algae, and microorganisms. Especially noteworthy is that the $\delta^{13}C$-ranges of C_3 and C_4 plants are virtually distinct and that the methanogenic bacteria exhibit an extremely large variation range.

One of the most important groups of all living matter is marine phytoplankton. Natural oceanic phytoplankton populations vary in $\delta^{13}C$-value by about 15‰ (Sackett et al. 1973; Wong and Sackett 1978). Rau et al. (1982) demonstrated that different latitudinal trends in the carbon isotope composition of plankton exist between the northern and the southern oceans: south of the equator the correlation between latitude and plankton ^{13}C content is significant, whereas a much weaker relationship exists in the northern oceans.

The unusual ^{13}C depletion in high-latitude southern ocean plankton has been puzzling for years and has been resolved after Rau et al. (1989, 1992) found a significant inverse relationship between high-latitude ^{13}C-depletion in plankton and the concentration of molecular CO_2 in surface waters. Because of the temperature dependence of CO_2 solubility in waters at equilibrium with atmospheric CO_2, a strong relationship between plankton $\delta^{13}C$ and CO_2 concentration may explain the apparent influence of temperature on plankton $\delta^{13}C$. It may also explain how similar low plankton $\delta^{13}C$-values can develop under what are apparently very different temperature and atmospheric CO_2 regimes: in modern cold high-southern latitude oceans under relatively low atmospheric CO_2 concentrations versus warm Cretaceous seas with high atmospheric CO_2 concentrations (Arthur et al. 1985).

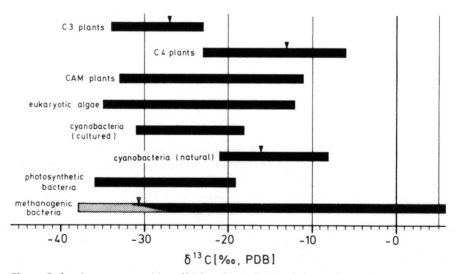

Fig. 56. Carbon isotope composition of higher plants, algae, and autotrophic prokaryotes. Means for some groups are indicated by *triangles*. (Schidlowski et al. 1983)

3.10.1.2
Chemical Components of Plant Material

A number of investigators have studied the carbon isotope composition of the major biochemical constituents of plants (Abelson and Hoering 1961; Degens et al. 1968a,b; Park and Epstein 1960; Parker 1964; Smith and Epstein 1970; DeNiro and Epstein 1977). These studies demonstrated that differences in ^{13}C contents exist between different chemical components of plants: sugar, cellulose, and hemicellulose exhibit $\delta^{13}C$-values close to the mean plant carbon isotopic composition, whereas pectin appears to be enriched in ^{13}C relative to the total plant. Lignin and lipids are depleted in ^{13}C relative to the total plant. The depletion of ^{13}C in the lipid fraction may be the main resason why oils characteristically have lower $\delta^{13}C$-values than the bulk organic matter in sediments.

Besides determining major organic constituents, the $\delta^{13}C$-value of individual organic components may represent a unique signature of its origin and its possible diagenetic alteration. Initial work by Abelson and Hoering (1961) involved the separation and analysis of individual amino acids. This study led to important insights on natural biosynthetic pathways and the effects of decarboxylation. The number of applications was limited, however, because conventional mass spectrometers usually need quantities of about 1 mg and the separation process is extremely labor intensive and may cause isotopic fractionations.

In recent years, significant progress has occurred through the commercial availability of a combined gas chromatography, combustion furnace, and mass spectrometry system which enables the analysis of individual components within complex organic mixtures. This analytical approach opens new perspectives for the study of biosynthetic pathways and processes of organic preservation and diagenetic alteration.

3.10.2
Isotope Fractionations of Other Elements During Photosynthesis

3.10.2.1
Hydrogen

During photosynthesis plants remove hydrogen from water and transfer it to organic compounds. As is indicated by the considerable H-isotope heterogeneity within components of living and dead biomass (Estep and Hoering 1980), different processes should be active in the assimilation of hydrogen into organic matter. Because plants utilize environmental water in photosynthesis, δD-values of plants are primarily determined by the δD-value of the water available for plant growth. Hydrogen enters the plant as water from roots in the case of terrestrial plants or via diffusion in the case of aquatic plants. In both cases, the water enters the organisms without any apparent fractionation. In higher terrestrial plants water transpires from the leaf due to evaporation, which is associated with an H-isotope fractionation up to 40–50‰ (White 1989).

Large negative isotope fractionations occur in biochemical reactions during the synthesis of organic compounds (Smith and Epstein 1970; Schiegl and Vogel

1970). A generalized picture of the hydrogen isotope fractionations in the metabolic pathway of plants is shown in Fig. 57 (after White 1989). There are systematic differences in the D/H ratios among classes of compounds in plants: lipids usually contain less deuterium than the protein and the carbohydrate fractions (Hoering 1975; Estep and Hoering 1980). The component typically analyzed in plants is cellulose, which is the major structural carbohydrate in plants (Epstein et al. 1976, 1977). Cellulose contains 70% carbon-bound hydrogen, which is isotopically nonexchangeable and 30% of exchangeable hydrogen in the form of hydroxyl groups (Epstein et al. 1976; Yapp and Epstein 1982). The hydroxyl-hydrogen readily exchanges with the environmental water and its D/H ratio is not a useful indicator of the D/H ratio of the water used by the plants.

This complex situation makes the interpretation of hydrogen isotope ratios in plants complicated and even more difficult for fossil organic matter. Smith et al. (1982) have demonstrated that δD-values of fossil coals and kerogens vary widely in the range –160–70‰. Until there is an improved understanding of the controlling factors, caution must be exercised in the interpretation of D/H ratios of organic materials.

3.10.2.2
Oxygen

The experimental difficulties in determining the oxygen isotope composition of biological materials lie in the rapid exchange between organically bound oxygen – in particular the oxygen of carbonyl and carboxyl functional groups – with wa-

Fig. 57. Generalized scheme of changes in hydrogen isotope composition in plants. (White 1989)

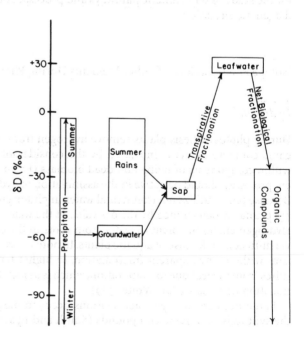

ter. This explains why studies on the oxygen isotope fractionation within living systems have concentrated on cellulose, the oxygen of which is only very slowly exchangeable (Epstein et al. 1977; DeNiro and Epstein 1979, 1981).

Oxygen potentially may enter organic matter from three different sources: CO_2, H_2O, and O_2. DeNiro and Epstein (1979) have shown that ^{18}O contents of cellulose for two sets of plant grown with water having similar oxygen isotope ratios, but with CO_2 having different oxygen isotope ratios, did not differ significantly. This means that CO_2 is in oxygen isotope equilibrium with the water. Similarly any isotopic labeling from molecular oxygen should be lost by exchange with water. Therefore, the isotopic composition of water determines the oxygen isotope composition of organically bound oxygen. Similar to hydrogen, oxygen isotope fractionation does not occur during uptake of soil water through the root, but rather in the leaf because of evapotranspiration.

3.10.2.3
Nitrogen

There are various pathways by which inorganic nitrogen can be fixed into organic matter during photosynthesis. Isotope fractionation will occur when the inorganic nitrogen source is in excess (Fogel and Cifuentes 1993). Ammonium can be assimilated in living matter by an enzymatic fixation. The oxidized forms nitrate and nitrite are reduced initially to ammonia, which, in turn, is then fixed into organic matter. Isotope fractionations during assimilation of NH_4 by algae varied extensively from -27 to $0‰$ (Fogel and Cifuentes 1993). A similar range of fractionations has been observed with algae grown on nitrate as the source of nitrogen.

3.10.2.4
Sulfur

Sulfur occurs mainly in proteins that typically display a C/S ratio of about 50. The processes responsible for the direct primary production of organically combined sulfur are the direct assimilation of sulfate by living plants and microbiological assimilatory processes in which organic sulfur compounds are synthesized.

At present, only a limited number of sulfur isotope measurements of biological materials are available. Mekhtiyeva and Pankina (1968) and Mekhtiyeva et al. (1976) have demonstrated that $^{34}S/^{32}S$ ratios of aquatic plants from a given water are slightly lower than for the sulfur of the dissolved sulfate. The same relationship has been obtained by Kaplan et al. (1963) for marine organisms (plants and animals).

3.10.3
Stable Isotopes as Indicators of Diet and Metabolism

A similarity in $\delta^{13}C$-values between animals and plants from the same environment was first noted by Craig as early as 1953. Later, many field and laboratory studies have documented small shifts of $1-2‰$ in ^{13}C and even smaller shifts in ^{34}S

between an organism and its food source (DeNiro and Epstein 1978; Peterson and Fry 1987; Fry 1988).

This technique has been widely used in tracing the origin of carbon, sulfur, and nitrogen in modern and prehistoric food webs (e.g., DeNiro and Epstein 1978) and culminates in the classic statement "You are what you eat plus/minus a few per mill." The precise magnitude of the isotopic difference between the diet and a particular tissue depends on the extent to which the heavy isotope is incorporated or lost during synthesis. In contrast to carbon and sulfur, nitrogen shows a 3–4‰ enrichment in ^{15}N in the muscle tissue, bone collagen, or whole organism relative to the food source (Minigawa and Wada 1984; Schoeninger and DeNiro 1984). When this fractionation is taken into account, nitrogen isotopes are also a good indicator of dietary source. In addition, the 3–4‰ $\delta^{15}N$ shift occurs with each trophic level along the food chain and thus provides a basis for establishing trophic structure.

The use of nitrogen isotopes in conjunction with carbon allows the distinction between terrestrial and marine food sources (Schoeninger and DeNiro 1984).

Archaeological studies have used the stable isotope analysis of collagen extracted from fossil bones to reconstruct the diet of prehistoric human populations (e.g., Schwarcz et al. 1985). Just as in modern food webs, carbon isotope values differentiate between prehistoric consumption of C_3 and C_4 plants and both ^{13}C and ^{15}N distinguish between marine and terrestrial food sources.

3.10.4
Recent Organic Matter

Typically, only a few percent of the initially biosynthesized organic matter escapes remineralization and becomes buried in sediments (Meyers 1994). Early diagenesis begins in the photic zone of oceans and lakes, continues during the sinking of particles, and is intense in the bioturbated surface layer of sediments. Despite the extensive losses of organic matter which occur during early diagenesis, the $^{13}C/^{12}C$ ratio appears to undergo little change. Thus, it is often possible to detect the origins of particulate organic carbon and organic matter in recent sediments via ^{13}C analysis. C_3-dominated terrestrial plant material has a $\delta^{13}C$-value of around –25‰ and organic matter of marine origin (phytoplankton) has a $\delta^{13}C$-value of around –20‰. Samples collected along riverine-offshore transects reveal very consistent and similar patterns of isotopic change from terrestrial to marine values. For example, in the Gulf of Mexico (Sackett and Thompson 1963, and others), in the St. Lawrence estuary (Tan and Strain 1983), in several North Sea estuaries (Salomons and Mook 1981), and in several other river deltas (e.g., Kennicutt et al. 1987), it is evident that the C-isotope signal of terrestrial organic matter derived from the land decreases with increasing proximity to the sea.

3.10.5
Fossil Organic Matter in Sediments

Organic matter in the bio- and geosphere is a complex mixture of living organisms and detrital remains. This complexity results from the multitude of source organisms, variable biosynthetic pathways, and transformations that occur during diagenesis and catagenesis.

Immediately after burial of the biological organic material into sediments, complex diagenetic changes occur in the organic matter. The biopolymers (e.g., polysaccharides and proteins) are attacked by microorganisms and are partly broken down to soluble components, while other parts polymerize and react to high-molecular-weight polycondensation products (i.e., humic substances). Carbon isotope shifts of a few per mill are connected with these diagenetic changes. They include isotope effects during bacterial degradation of the biopolymers, which preferentially eliminate ^{13}C-enriched carbohydrates and proteins and preserve ^{12}C-enriched lipids. Decarboxylation reactions remove ^{13}C-enriched carboxyl groups, leading to ^{13}C depletion in the residual.

Considered as a whole, recent marine sediments show a mean $\delta^{13}C$-value of $-25‰$ (Deines 1980b). Some ^{13}C loss occurs with transformation to kerogen, leading to an average $\delta^{13}C$-value of $-27.5‰$ (Hayes et al. 1983). This ^{13}C depletion might be best explained by the large losses of CO_2 that occur during the transformation to kerogen and which are especially pronounced during the decarboxylation of some ^{13}C-rich carboxyl groups. With further thermal maturation, the opposite effect of a ^{13}C enrichment is observed. Experimental studies of Chung and Sackett (1979), Peters et al. (1981), and Lewan (1983) indicate that thermal alteration produces a maximum ^{13}C change of about $+2‰$ in kerogens. Changes of more than $2‰$ are most probably not due to isotope fractionation during thermal degradation of kerogen, but rather to isotope exchange reactions between kerogen and carbonates.

As previously noted, recent marine organic carbon is ^{13}C enriched relative to terrestrial organic carbon. This distinction has been used to differentiate between these two sources in sediments (Brown et al. 1972). There is, however, increasing evidence that the carbon isotope fractionation associated with the production of marine organic matter has changed with geological time, while that associated with the production of terrestrial organic matter has been nearly constant (Arthur et al. 1985; Hayes et al. 1989; Popp et al. 1989; Whittacker and Kyser 1990). Popp et al. (1989) proposed that the decreased isotopic fractionation between marine carbonates and organic matter from Early to Mid-Cenozoic may record variations in the abundance of atmospheric CO_2. Thus, environmental changes may have affected marine phytoplankton to the extent that they developed an alternate metabolism with a resultant increase in $\delta^{13}C$-values.

The recently developed technique to measure individual organic molecules demonstrates that isotope variations within the individual components of sedimentary organic matter by far exceed those observed between samples of total organic matter (Freeman et al. 1990; Kenig et al. 1994, and others). In the Messel Shale, Freeman et al. (1990) observed C-isotope variations from -73.4 to $-20.9‰$. This extremely large range of $\delta^{13}C$-values can be interpreted as representing a mixture of secondary (probably bacterially mediated) processes and primary producers. This is consistent with results by Kenig et al. (1994), who studied the chlorophyll-derived alkanes phytane and pristane in the Jurassic Oxford Clay formation and demonstrated that the bulk organic matter is extensively "reworked" by heterotrophic organisms.

In a study of individual hydrocarbons extracted from sedimentary rocks of Proterozoic age, Logan et al. (1995) observed a ^{13}C enrichment relative to the kerogen, which is opposite to the trend generally found. This relationship may indicate that hydrocarbons of Proterozoic age are derived mainly from bacteria or other heterotrophs rather than photosynthetic organisms and that, in turn, a transition from anaerobic to aerobic conditions of the ocean occurred across the Precambrian–Cambrian boundary.

3.10.6
Oil

The study of crude oils and natural gases, combined with stable isotope analysis (^{13}C, D, ^{34}S, ^{15}N), has become a powerful tool in petroleum exploration (Fuex 1977; Stahl 1977; Schoell 1984a,b; Sofer 1984). The isotopic composition of crude oil is mainly determined by the isotopic composition of its source material, more specifically the type of kerogen and the sedimentary environment in which it has been formed. Secondary effects such as biodegradation, water washing, and migration distances appear to have only minor effects on its isotopic composition.

Variation in ^{13}C has been the most widely used parameter. Generally, oils are depleted by 1–3‰ compared to their source rocks, and the various chemical compounds within crude oils show small, but characteristic $\delta^{13}C$-differences. With increasing polarity the ^{13}C content increases from the saturated to aromatic hydrocarbons to the heterocomponents (N, S, O compounds) and to the asphaltene fraction. These characteristic differences in ^{13}C have been used for correlation purposes. Sofer (1984) plotted the ^{13}C contents of the saturate and aromatic fractions against

Fig. 58. "Petroleum-type curves" of different oil components from the North Sea. Diagram shows a positive oil-oil correlation and a negative source rock-oil correlation. SAT, saturated hydrocarbons; AROM, aromatic hydrocarbons; NSOs, heterocomponents; ASPH, asphaltenes. (Stahl 1977)

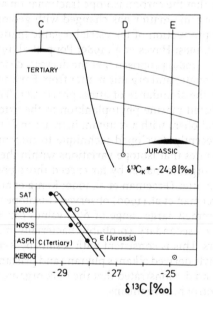

each other. Oils and suspected source rock extracts that are derived from similar types of source materials will plot together in such a graph whereas those derived from different types of source material will plot in other regions of the graph. The approach of Stahl (1977) and Schoell (1984) is somewhat different: the ^{13}C contents of the different fractions are plotted as shown in Fig. 58. In this situation, oils derived from the same source rock will define a near linear relationship in the plot. Figure 58 illustrates a positive oil-oil correlation and a negative oil-source rock correlation.

A separate analysis of the isoprenoid-hydrocarbons, pristane, and phytane, for which a common origin from chlorophyll had been generally assumed, may indicate that these two components have different origins (Freeman et al. 1990). Other classes of biomarkers, such as the hopanes, are also not always derived from a common precursor (Schoell et al. 1992). These results indicate that the origin and fate of organic compounds are far more complicated than was previously assumed.

3.10.7
Coal

Carbon and hydrogen isotope compositions of coals are rather variable (Schiegl and Vogel 1970; Redding et al. 1980; Smith et al. 1982). Different plant communities and climates may account for these variations. Due to the fact that, during coalification, the amount of methane and other higher hydrocarbons liberated is small compared to the total carbon reservoir, very little change in carbon isotope composition seems to occur with increasing grade of coalification. With respect to hydrogen the reservoir is smaller, which may explain why differences in D contents up to 50‰ have been observed (Redding et al. 1980).

The origin and distribution of sulfur in coals is of special significance, because of the problems associated with the combustion of coals. Sulfur in coals usually occurs in different forms, as pyrite, organic sulfur, sulfates, and elemental sulfur. Pyrite and organic sulfur are the most abundant forms. Organic sulfur is primarily derived from two sources: the original organically bound plant sulfur preserved during the coalification process and biogenic sulfides which reacted with organic compounds during the biochemical alteration of plant debris.

Studies by Smith and Batts (1974), Smith et al. (1982), Price and Shieh (1979), and Hackley and Anderson (1986) have shown that organic sulfur exhibits rather characteristic S-isotope variations which correlate with sulfur contents. In low-sulfur coals $\delta^{34}S$-values of organic sulfur are rather homogeneous and reflect the primary plant sulfur. By contrast, high-sulfur coals are isotopically more variable and typically have more negative $\delta^{34}S$-values, suggesting a significant contribution from bacteriogenic sulfides.

3.10.8
Natural Gas

Natural gas has been found in a wide variety of environments. While methane is always a major constituent of the gas, other components may be higher hydrocarbons (ethane, propane, butane), CO_2, H_2S, N_2, and rare gases. Two different types of gas occurrences can be distinguished – biogenic and thermogenic gas – the most useful parameters in distinguishing both types being their $^{13}C/^{12}C$ and D/H ratios. Complications in assessing sources of natural gases are introduced by mixing, migration, and oxidative alteration processes. For practical application an accurate assessment of the origin of a gas, the maturity of the source rock, and the timing of gas formation would be desirable. During the past 20 years a variety of models have been published that describe carbon and hydrogen isotope variations of natural gases (Berner et al. 1995; Galimov 1988; James 1983, 1990; Rooney et al. 1995; Schoell 1983, 1988; Stahl and Carey 1975).

Rather than using the isotopic composition of methane alone, James (1983, 1990) and others have demonstrated that carbon isotope fractionations between the hydrocarbon components (particularly propane, iso-butane and normal butane) within a natural gas can be used with distinct advantages to determine maturity, gas-source rock, and gas-gas correlations. With increasing molecular weight, from C_1 to C_4, a ^{13}C enrichment is observed which approaches the carbon isotope composition of the source (Fig. 59). Another advantage of using the carbon isotope composition of higher hydrocarbons is the $^{13}C/^{12}C$ ratio is generally unaffected by migration.

Fig. 59. $\delta^{13}C$ of individual hydrocarbon gases (C_1–C_4) from three different gas fields of western Canada. (James 1990; reprinted with permission)

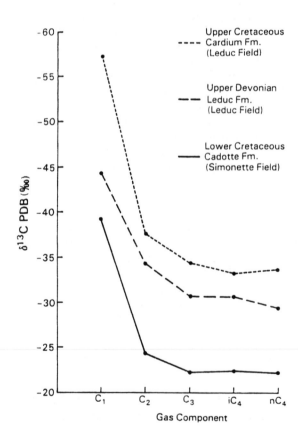

3.10.8.1
Biogenic Gas

According to Rice and Claypool (1981), over 20% of the world's natural gas accumulations are of biogenic origin. Biogenic methane commonly occurs in recent anoxic sediments and is well documented in both freshwater environments, such as lakes and swamps, and in marine environments, such as estuaries and shelf regions. Two primary metabolic pathways are generally recognized for methanogenesis: fermentation of acetate and reduction of CO_2. Although both pathways may occur in both marine and freshwater environments, CO_2 reduction is dominant in the sulfate-free zone of marine sediments, while acetate fermentation is dominant in freshwater sediments.

During microbial action, kinetic isotope fractionations on the organic material by methanogenic bacteria result in methane very much depleted in ^{13}C, typically with $\delta^{13}C$-values between –110 and –50‰ (Schoell 1984, 1988; Rice and Claypool 1981; Whiticar et al. 1986). In marine sediments, methane formed by CO_2 reduction is often more depleted in ^{13}C than methane formed by acetate fermentation in freshwater sediments. Thus, typical $\delta^{13}C$ ranges for marine sediments are between –110 and –60‰, while those for methane from freshwater sediments are from –65 to –50‰ (Whiticar et al. 1986).

The difference in composition between methane of freshwater and of marine origin is even more pronounced on the basis of hydrogen isotopes. Marine bacterial methane has δD-values between –250 and –170‰ while biogenic methane in freshwater sediments is strongly depleted in D, with δD-values between –400‰ and –250‰ (Whiticar et al. 1986). Different hydrogen sources may account for these large differences: formation waters supply the hydrogen during CO_2 reduction, whereas during fermentation some three-fourths of the hydrogen comes directly from the methyl group, being extremely depleted in D.

3.10.8.2
Thermogenic Gas

Thermogenic gas is produced when organic matter is buried to depth. Increasing temperatures modify the organic matter due to various chemical reactions, such as cracking and hydrogen diproportionation in the kerogen. ^{12}C–^{12}C bonds are preferentially broken during the first stages of the maturation of organic matter. As this results in an ^{13}C enrichment of the residue, more ^{13}C–^{12}C bonds are broken with increasing temperatures, which produces higher $\delta^{13}C$-values. Thermal cracking experiments carried out by Sackett (1978) have confirmed this process and demonstrated that the resulting methane is depleted in ^{13}C by some 4 –25‰ relative to the parent material. Thus, thermogenic gas typically has $\delta^{13}C$-values between –50‰ and –20‰ (Schoell 1980, 1988). Gases generated from nonmarine (humic) source rocks are isotopically enriched relative to those generated from marine (sapropelic) source rocks at equivalent levels of maturity. In contrast to $\delta^{13}C$-values, δD-values are independent of the composition of the precursor material, but solely depend on the maturity of kerogen.

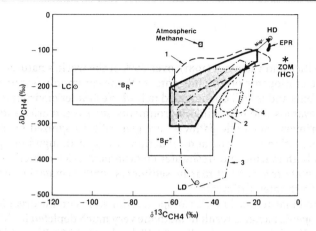

Fig. 60. $\delta^{13}C$ and δD variations of natural gases of different origins. *Fields B_R and B_F represent bacterial methanes that form by CO_2 reduction and fermentation, respectively. Heavy outlined area encompasses methane of thermogenic origin, wherein the shaded part depicts methane associated with oils. Fields with numbers are methanes from specific areas.* 1 Sacramento Basin; 2 Cooper Basin, Western Australia; 3 Canadian Shield gases; 4 geothermal methane. LC, HC and LD, HD are the highest and lowest concentrations for ^{13}C and D, respectively, found so far in natural methanes. (After Schoell 1988; Reprinted from Chemical Geology, Vol. 71, Multiple origins of methane in the earth, pp 1-10, © 1988, with kind permission from Elsevier Science – NL, Sara Burgerhartstraat 25, 1055 KV Amsterdam, The Netherlands)

In conclusion, the combination of carbon and hydrogen isotope analysis of natural gases is a powerful tool to discriminate different origins of gases. In a plot of $\delta^{13}C$ versus δD (see Fig. 60), not only is a distinction of biogenic and thermogenic gases from different environments clear, but it is also possible to delineate mixtures between the different types.

Nitrogen is sometimes a major constituent of natural gases, and the origin of this nitrogen is still enigmatic. While a certain fraction is released from sedimentary organic matter during burial, several nonsedimentary sources of nitrogen may also contribute to the natural gas. Boigk et al. (1976) postulated that $\delta^{15}N$-values like $\delta^{13}C$-values may indicate relationships between hydrocarbon gas deposits and their potential source rocks. By analyzing nitrogen-rich natural gases from the California Great Valley, Jenden et al. (1988) demonstrated, however, that these gases had a complex origin involving mixing of multiple sources. These authors interpreted relatively constant $\delta^{15}N$-values between 0.9‰ and 3.5‰ as indicating a deep-crustal metasedimentary origin. Hydrocarbon-rich and nitrogen-rich gases can be thus genetically unrelated.

3.10.8.3
Abiogenic Methane

Methane emanating in mid-ocean ridge hydrothermal systems is one of the few occurrences for which an abiogenic formation can be postulated with confidence. $\delta^{13}C$-values between –18‰ and –15‰ (Welhan 1988) are outside the range of most biogenic methanes. This C-isotope composition alone does not unambiguously argue for an abiogenic origin, because high-temperature equilibration processes between CO_2 and CH_4 may account for a ^{13}C enrichment in the methane. The most enriched methane so far known has a $\delta^{13}C$-value of $-7‰$ and is derived from seeps in the Zambales Ophiolite in the Philippines (Abrajano et al. 1988). The classification of methane being of abiogenic origin remains, nevertheless, problematic.

3.11
Sedimentary Rocks

Sediments are the weathering products and residues of magmatic, metamorphic, and sedimentary rocks after transport and accumulation in water and air. As a result, sediments may be complex mixtures of material that has been derived from multiple sources. It is customary to consider sedimentary rocks, and the components of sedimentary rocks, in two categories: clastic and chemical. Transported fragmental debris of all kinds makes up the clastic component of the rock. Inorganic and organic precipitates from water belong to the chemical constituents. According to their very different modes and low temperatures of formation, sedimentary rocks may be extremely variable in isotopic composition. For example, the $\delta^{18}O$-values of sedimentary rocks span a large range from about $+10‰$ (certain sandstones) to about $+44‰$ (certain cherts).

3.11.1
Clay Minerals

By comparison with many other silicate minerals, isotope studies of natural clays are confronted with a number of special problems related to their small particle size and, hence, much larger specific surface area and presence of interlayer water in certain clays. Surfaces of clays are invariably associated with one- or two-layer-thick adsorbed water. Savin and Epstein (1970a) demonstrated that adsorbed and interlayer water exchange isotopically with atmospheric water vapor in times of hours. Complete removal for analysis of interlayer water, with the total absence of isotopic exchange between it and the hydroxyl group, may not be possible in all instances (Lawrence and Taylor 1971). Therefore, the precision of H- and O-isotope analyses of clays is often a little lower than that of most other minerals (Sheppard and Gilg 1995).

Savin and Epstein (1970a,b) and Lawrence and Taylor (1971) established the general isotope systematics of clay minerals from continental and oceanic environments. Subsequent reviews by Savin and Lee (1988) and Sheppard and Gilg (1995) have applied the isotope studies of clay minerals to a wide range of geological problems. All applications depend on the knowledge of isotope fractionation factors between clay minerals and water, the temperature, and the time when isotopic exchange with the clay ceased. Because clay minerals may be composed of a mixture of detrital and authigenic components, and because particles of different ages may have exchanged to varying degrees, the interpretation of isotopic variations of clay mineral data is rather complicated.

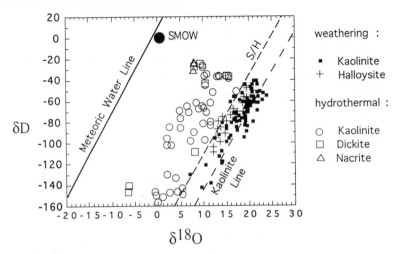

Fig. 61. Compilation of δD- and δ^{18}O-values for kaolinites and related minerals from weathering and hydrothermal environments. The Meteoric Water Line, kaolinite weathering line, and supergene/hypogene (*S/H*) line are given for reference. (After Sheppard and Gilg 1995)

Do natural clay minerals retain their initial isotopic compositions? Evidence concerning the extent of isotopic exchange for natural systems is contradictory (Sheppard and Gilg 1995). Many clay minerals such as kaolinite, smectite, and illite are often out of equilibrium with present-day local waters. This is not to imply that these clay minerals never undergo any post-formational or retrograde exchange. Sheppard and Gilg (1995) concluded that convincing evidence for complete O- and/or H-isotope exchange without recrystallization is usually lacking, unless the clay has been subjected to either higher temperatures or an unusual set of geological circumstances. Thus, clay minerals formed in contact with meteoric waters should depend on the Meteoric Water Line relationship and plot on lines which are subparallel to the Meteoric Water Line. This implies that some information of past environments is usually recorded in clay minerals (see Fig. 61). In a survey of ancient kaolinites from North America, Lawrence and Raskes-Meaux (1993) concluded that ^{18}O/^{16}O ratios have been preserved, whereas D/H ratios have been partially re-equilibrated.

3.11.2
Clastic Sedimentary Rocks

Clastic sedimentary rocks are composed of detrital grains that normally retain oxygen isotope composition of their source and of authigenic minerals formed during weathering and diagenesis, whose isotopic composition is determined by the physicochemical environment in which they formed. This means authigenic minerals formed at low temperatures will be enriched in ^{18}O compared to detrital minerals of igneous origin (Savin and Epstein 1970c). Due to the difficulty of separating authigenic overgrowths from detrital cores, only few studies of this kind have been reported in the literature. One such example is that of Girard et al. (1988), the results of which are shown in Fig. 62. These authors were able to concentrate K-feldspar overgrowths from arkoses, which permitted extrapolation to the pure authigenic and detrital end-members whose respective $\delta^{18}O$-values were estimated to be 20.2 and 9.1‰. This detrital component probably originates from the granitic basement.

How ^{18}O enriched the authigenic mineral will be is determined by fluid composition, temperature, and the effective mineral/water ratio. Should the fluid be a low-^{18}O meteoric water, the oxygen isotope composition of the precipitating mineral will be lower, given no change in temperature (Longstaffe 1989). Thus, the changes that occur in sedimentary rocks during diagenesis are largely a function of fluid composition and temperature.

Determination of the temperature to which a sedimentary unit has been heated during various stages of diagenesis is an important objective of geothermometry. One way to estimate temperatures employs the oxygen isotope composition of diagenetic assemblages. For example, using quartz–illite pairs from the Precambrian Belt Supergroup, Eslinger and Savin (1973) calculated temperatures that range from 225 to 310 °C, with increasing depth, $\delta^{18}O$-values that are consistent with the observed mineralogy and are reasonable for the grade of burial metamorphism. This approach assumes that the diagenetic minerals used have isotopically equilibrated with each other and that no retrograde re-equilibration occurred following maximum burial.

Another application of stable isotopes in clastic rocks is the analysis of weathering profiles, which can potentially provide insight into the continental climate during their formation. Despite this potential, only few studies (Bird and Chivas 1989; Bird et al. 1992) have used this approach because of (1) the imprecise knowledge of mineral–water fractionations at surficial temperatures and (2) the difficulty of obtaining pure phases from complex very fine grained rocks. Bird et al. (1992) devel-

Fig. 62. Relationship between $\delta^{18}O$-values of concentrates of authigenic K-feldspar overgrowths and volume percent of detrital K-feldspar in three arkoses from Angola. The concentrations of detrital feldspar were estimated by point counting based on cathodoluminescence. The extrapolated $\delta^{18}O$-values apply to the pure authigenic and detrital feldspar components. (After Faure 1992)

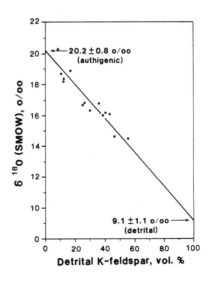

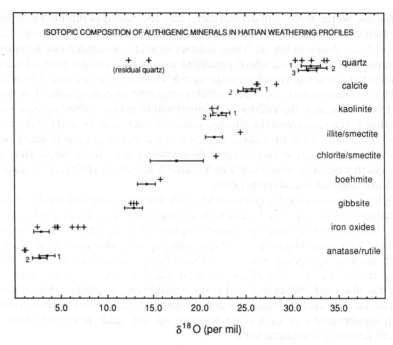

Fig. 63. Predicted (*bars*) and observed (*crosses*) oxygen isotope composition of separated minerals from Haitian weathering profiles. The ranges of predicted δ^{18}O-values were calculated assuming a temperature of 25 °C and a meteoric water δ^{18}O-value of –3.1‰. (After Bird et al. 1992; Reprinted from Geochimica et Cosmochimica Acta, Vol. 56 No. 7, Oxygen-isotope systematics in a multiphasse weathering system in Haiti, pp 2831-2838, © 1992, with kind permission from Elsevier Science Ltd, The Boulevard, Langford Lane, Kidlington OX5 1GB, UK)

oped partial dissolution techniques and used this methodology to separate nine pure minerals from a lateritic soil in Haiti (see Fig. 63). The measured δ^{18}O-values for some minerals agree well with ^{18}O/^{16}O ratios predicted from available fractionation factors, whereas other do not. Discrepancies might be due to incorrect fractionation factors for the respective minerals or to processes that may have influenced the formation of particular minerals (e.g., evaporation) (Bird et al. 1992).

3.11.3
Biogenic Silica and Cherts

Due to the large oxygen isotope fractionation between SiO_2 and water at low temperatures, biogenic silica and cherts have the highest ^{18}O/^{16}O ratios observed in rocks. As in carbonates, the oxygen isotope composition in biogenic silica is determined by temperature and the isotopic composition of water. Thus, the isotope composition of biogenic silica has a powerful potential as a geothermometer. However, the presence of loosely bound water within cherts and biogenic silica precipitates complicates the analytical determination of the silica oxygen. For example, partial exchange of silica with water occurs during cleaning procedures of biogenic silica samples. Approximately 10%–20% of the total silica oxygen appears to be involved in this exchange (Labeyrie and Juillet 1982). Using controlled exchange with waters of different isotope composition, Labeyrie and Juillet (1982) and Leclerc and Labeyrie (1987) were able to estimate the isotope ratio of both exchanged and unexchanged silica-bound oxygen.

An entirely different analytical approach was taken by Haimson and Knauth (1983) and Matheney and Knauth (1989), who utilized a stepwise fluorination technique. These authors noted that the first oxygen fractions were ^{18}O depleted compared with oxygen recovered in later fractions, suggesting that the water-rich components of hydrous silica react preferentially in the early steps of fluorination. Oxygen isotope fractionations determined by these two different analytical methods are consistent with each other. As recent investigations of Shemesh et al. (1992, 1995) have suggested, diatoms indeed can be used for reconstructions of ocean water temperatures. This conclusion was, however, questioned by Schmidt et al. (1996), who argued that successive isotope exchange reactions of diatomaceous silica with pore waters control the isotope composition of fossil diatoms in sediments.

In sediments opaline skeletons are frequently dissolved and opal-CT is precipitated. As was shown from the early studies from Degens and Epstein (1962), like carbonates, cherts exhibit temporal isotopic variations: the older cherts having lower ^{18}O contents. Thus, cherts of different geological ages may contain a record of temperature, isotopic composition of ocean water, and diagenetic history. There is still debate about which of these factors is the most important (see discussion on p. 117).

3.11.4
Marine Carbonates

3.11.4.1
Oxygen

In 1947, Urey presented a paper concerning the thermodynamics of isotopic systems and suggested that variations in the temperature of precipitation of calcium carbonate from water should lead to measurable variations in the $^{18}O/^{16}O$ ratio of the calcium carbonate. He postulated that the determination of temperatures of the ancient oceans should be possible, in principle, by measuring the ^{18}O content of fossil calcite shells. The first paleotemperature "scale" was introduced by McCrea (1950); subsequently this scale has been refined several times. Through experiments which compare the actual growth temperatures of foraminifera with calculated isotope temperatures, Erez and Luz (1983) determined the following temperature equation:

$$T \,°C = 17.0 - 4.52 \,(\,\delta^{18}O_c - \delta^{18}O_w) + 0.03 \,(\,\delta^{18}O_c - \delta^{18}O_w)^2$$

According to this equation, an ^{18}O increase of 0.26‰ represents a 1 °C temperature decrease. However, before a meaningful temperature calculation can be carried out several assumptions have to be fulfilled.

The isotopic composition of an aragonite or calcite shell will remain unchanged until the shell material dissolves and recrystallizes during diagenesis. In most shallow depositional systems, C- and O-isotope ratios of shells are fairly stable and resistant to diagenetic changes. With increasing depths of burial and time the chances of diagenetic effects generally increase. Because fluids contain much less carbon than oxygen, $\delta^{13}C$-values are thought to be less affected by diagenesis than $\delta^{18}O$-values. Criteria of how to prove primary preservation are not always clearly resolved (see discussion in Sect. 3.11.5).

Shell-secreting organisms to be used for paleotemperature studies have to have been precipitated in isotope equilibrium with ocean water. As was shown by studies of Weber and Raup (1966a,b) and Weber (1968), some organisms precipitate their skeletal carbonate in equilibrium with the water in which they live, but others do not. Wefer and Berger (1991) summarized the importance of the so-called "vital effect" on a broad range of organisms (see Fig. 64). As regards oxygen isotopes, most organisms precipitate $CaCO_3$ close to equilibrium, or if disequilibrium prevails, the isotopic difference from equilibrium remains constant. For carbon, disequilibrium is the rule, with $\delta^{13}C$-values much more negative than expected at equilibrium. As discussed below, this does not preclude the reconstruction of the $^{13}C/^{12}C$ ratio of the palaeo-ocean waters.

Isotopic disequilibria effects can be classified as either metabolic or kinetic (McConnaughey 1989a,b). Metabolic isotope effects apparently result from changes in the isotopic composition of dissolved inorganic carbon in the neighborhood of the precipitating carbonate caused by photosynthesis and respiration. Kinetic isotope effects result from discrimination against ^{13}C and ^{18}O during hydration and hydroxylation of CO_2. Strong kinetic disequilibrium behavior appears to be often associated with rapid calcification (McConnaughey 1989b).

Most oceanic paleoclimate studies have concentrated on foraminifera. Since the first pioneering paper of Emiliani (1955), numerous cores from various sites of the DSDP and ODP program have been analyzed and, when correlated accurately, have produced a well-established oxygen isotope curve for the Pleistocene and for the Tertiary. These core studies have demonstrated that similar $\delta^{18}O$-variations are observed in all areas. With independently dated time scales on hand, these systematic ^{18}O variations result in synchronous isotope signals in the sedimentary record because the mixing time of the oceans is relatively short (10^3 years). These signals provide stratigraphic markers, enabling correlations between cores which may be thousands of kilometers apart. Several Pleistocene biostratigraphic datums have been calibrated with oxygen isotope stratigraphy, which helps to confirm their synchrony. This correlation has greatly facilitated the recognition of both short and long time periods of characteristic isotopic composition, and times of rapid changes from one isotope state to another, thus making oxygen isotope stratigraphy a practical tool in modern paleoceanographic studies. Figure 65 shows the oxygen isotope curve for the Pleistocene. This diagram exhibits several striking features: the most obvious one is the cyclicity; furthermore fluctuations never go beyond a certain maximum value on either side of the range. This seems to imply that very effective feedback mechanisms are at work stopping the cooling and warming trends at some maximum level. The "sawtooth"-like curve in Fig. 65 is characterized by very steep gradients: maximum warm periods are immediately followed by maximum cold periods.

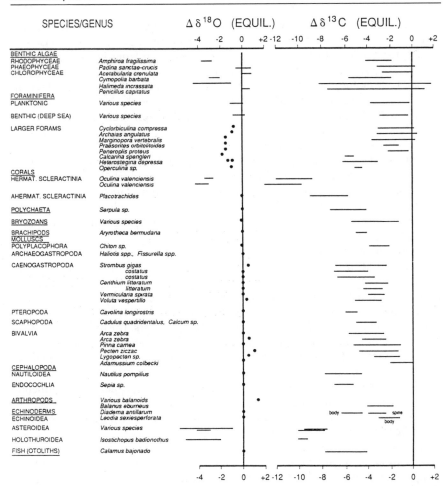

Fig. 64. $\delta^{18}O$ and $\delta^{13}C$ differences (Δ-values) from equilibrium isotope composition of extant calcareous species. (After Wefer and Berger 1991; Reprinted from Marine Geology, Vol. 100, Isotope paleontology: growth and composition of extant calcareous species, pp 207-248, © 1991, with kind permission from Elsevier Science – NL, Sara Burgerhartstraat 25, 1055 KV Amsterdam, The Netherlands)

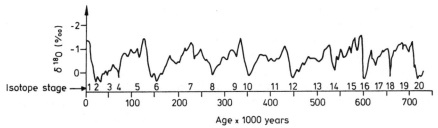

Fig. 65. Composite $\delta^{18}O$-fluctuations in the foraminifera species *G. sacculifer* from Caribbean cores. (Emiliani 1978)

Emiliani (1955) introduced the concept of "isotopic stages" by designating stage numbers for identifiable events in the marine foraminiferal oxygen isotope record for the Pleistocene. Odd numbers identify interglacial or interstadial (warm) stages, whereas even numbers define ^{18}O-enriched glacial (cold) stages. A second terminology used for subdividing isotope records is the concept of terminations, which was introduced by Broecker and Van Donk (1970). Terminations labeled with Roman numbers I, II, III, etc. describe rapid transitions from peak glacial to peak interglacial values.

A careful examination of the curve shown in Fig. 65 shows a periodicity of approximately 100 000 years. Hays et al. (1976) argued that the main structure of the oxygen isotope record is caused by variations in solar insolation, promoted by variations in the Earth's orbital parameters. Thus, isotope data have played a capital role in the confirmation of the "Milankovitch Theory," which argues that the isotope record is a response to the forcing of the orbital parameters operating at specific frequencies.

Besides reflecting oceanic temperatures, a variable isotopic composition of the ocean is another factor responsible for ^{18}O variations in foraminifera. A crucial control is the salinity: ocean water with salinities greater than 3.5% has a higher ^{18}O content, because ^{18}O is preferentially depleted in the vapor phase during evaporation, whereas water with salinities lower than 3.5% has a lower ^{18}O content because it has been diluted by fresh waters, especially ice meltwaters. The other factor which causes variations in the isotopic composition of ocean water is the volume of low-^{18}O ice present on the continents. As water is removed from the ocean during glacial periods, and temporarily stored on the continents as ^{18}O-depleted ice, the $^{18}O/^{16}O$ ratio of the global ocean increases in direct proportion to the volume of continental glaciers. The magnitude of the temperature effect versus the ice volume effect can be largely resolved by separately analyzing planktonic and benthic foraminifera. It might be expected that the temperature of deep-water masses is more or less constant, as long as ice caps exist at the Poles. Thus, the oxygen isotope composition of benthic dwelling organisms should preferentially reflect the change in the isotopic composition of the water, while the $\delta^{18}O$-values of planktonic foraminifera should be affected by both temperature and isotopic water composition.

The best approach to disentangle the effect of ice volume and temperature is by the study of areas where constant temperatures have prevailed for long periods of time, such as the western tropical Pacific Ocean or the tropical Indian Ocean. At the other end of the temperature spectrum is the Norwegian Sea, where the deep water temperature is near freezing point today and, therefore, could not have been significantly lower during glacial time. Within the framework of this set of limited assumptions, a reference record of the ^{18}O variations of a water mass which has experienced no temperature variations during the last climatic cycle can be obtained (Labeyrie et al. 1987).

It is also known from the investigations of coral reefs that during the last glacial time sea level was lowered by 125 m. Fairbanks (1989) calculated that the ocean was enriched by 1.25‰ during the last glacial maximum, which indicates an ^{18}O difference of 0.1‰ for a 10-m increase in sea level. This relationship is obviously valid only for the last glacial period, because thicker ice shields might concentrate ^{16}O even more than smaller ones.

Fig. 66. $\delta^{18}O$-values of benthic foraminifera from Deep Sea Drilling Program sites spanning the past 70 Ma. The long-term increase in $\delta^{18}O$-values reflects cooling of the deep ocean and growth of ice sheets at high latitudes. (After Raymo and Ruddiman 1992; Reprinted with permission from Nature, Vol. 359 No. 6391, © 1992, Macmillan Magazines Limited)

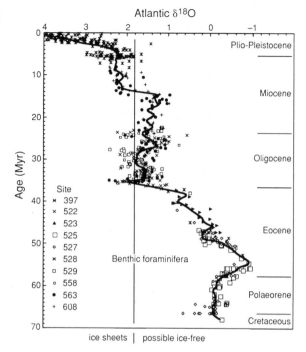

Figure 66 traces the oxygen isotope record back through the last 70 Ma, which clearly shows a global cooling throughout the Tertiary. The sharp increase in isotopic values in the early Oligocene may reflect the first major ice-growth event on Antarctica, which followed roughly 20 million years of cooling. Further $\delta^{18}O$ increases in the middle Miocene and late Pliocene represent subsequent increases in ice volume in Antarctica and in the Northern Hemisphere, respectively, although the exact portion attributable to ice volume rather than deep ocean cooling remains uncertain (Raymo and Ruddiman 1992).

3.11.4.2
Carbon

A large number of studies have investigated the use of ^{13}C contents of foraminifera as a paleo-oceanographic tracer. As previously noted, $\delta^{13}C$-values are not in equilibrium with seawater; however, by assuming that disequilibrium $^{13}C/^{12}C$ ratios are invariant with time, on average, then systematic variations in C-isotope composition may reflect variations in ^{13}C contents of ocean water. The first record of carbon isotope compositions in Cenozoic deep-sea carbonates was given by Shackleton and Kennett (1975). They clearly demonstrated that planktonic and benthic foraminifera yield consistent differences in $\delta^{13}C$-values, the former being enriched in ^{13}C by about 1‰ relative to the latter. This ^{13}C enrichment in plankton-

Fig. 67. δ^{13}C-values of benthic foraminifera species. The δ^{13}C-value for the dissolved bicarbonate in deep equatorial water is shown by the *vertical line*. (After Wefer and Berger 1991; Reprinted from Marine Geology, Vol. 100, Isotope paleontology: growth and composition of extant calcareous species, pp 207-248, © 1991, with kind permission from Elsevier Science – NL, Sara Burgerhartstraat 25, 1055 KV Amsterdam, The Netherlands)

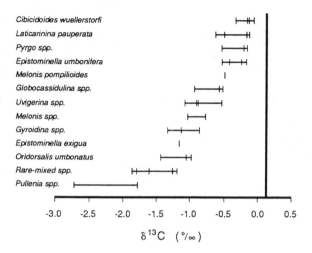

ic foraminifera is due to photosynthesis which removes ^{12}C preferentially from the surface layer into organic matter. A portion of this organic matter settles into deep water, where it is reoxidized, which causes a ^{12}C enrichment in the deeper water masses. Figure 67 presents δ^{13}C-values of benthic foraminifera ranked by their relative tendency to concentrate ^{13}C. δ^{13}C-values in planktonic and benthic foraminifera can be used to monitor CO_2 variations in the atmosphere by measuring the vertical carbon isotope gradient, which is a function of the biological carbon pump. This approach was pioneered by Shackleton et al. (1983), who showed that enhanced contrast between surface waters and deeper waters was correlated with intervals of reduced atmospheric CO_2 contents. Increased organic carbon production in surface waters (possibly caused by enhanced nutrient availability) leads to removal of carbon from surface waters, which in turn draws CO_2 down from the atmospheric reservoir through re-equilibration.

Another application of carbon isotopes in foraminifera is to trace deep water circulation (Bender and Keigwin 1979; Duplessy et al. 1988). Since dissolved carbonates in the deep waters become isotopically lighter with time after leaving the surface in the area of their formation due to the increasing oxidation of organic material, comparison of sites of similar paleodepth in different areas can be used to trace the pathway by deep waters as they move from their sources. Such a reconstruction can be carried out by analyzing δ^{13}C-values of well-dated foraminifera.

3.11.5
Diagenesis of Limestones

Diagenetic modification of freshly deposited carbonates begins immediately after the formation of primary carbonates and can be divided into two major subsequent pathways, often termed as meteoric and burial diagenesis (Veizer 1992). These two pathways converge with time, although overlaps are possible even in the early stages of diagenesis.

3.11.5.1
Meteoric Pathway

Carbonate sediments deposited in shallow marine environments are often exposed to the influence of meteoric waters during later phases of their history. Such waters are generally charged with CO_2 and transform the original unstable mineral assemblage aragonite and Mg-calcite into a low-Mg calcitic limestone via dissolution–reprecipitation processes. The other important diagenetic process is the oxidation of organic matter. With burial, the sediment passes successively through different zones which are characterized by distinct redox reactions that are mediated by assemblages of specific bacteria. The usual isotopic signature of these processes will be a shift toward lighter C-isotope values, the degree of ^{13}C depletion being proportional to the relative contribution of carbon from the precursor carbonate and from the oxidation of organic matter. Only under special conditions of fermentation may the CO_2 released be isotopically heavy, which may cause a shift in the opposite direction.

3.11.5.2
Burial Pathway

This type of diagenetic stabilization is best documented in deep sea environments. Entrapped pore waters are of marine origin and in equilibrium with the assemblage of carbonate minerals. The conversion of sediment into limestone is not achieved by a chemical gradient, but rather through pressure and temperature rise due to deposition of younger sediments. In burial diagenesis, in contrast to the meteoric pathway, the fluid flow is confined to squeezing of pore waters upwards into the overlying water column. Theoretically, because the ^{18}O is of seawater origin, O-isotope ratios should not change appreciably with burial. Yet, with increasing depth, the deep-sea sediments and often also the pore waters exhibit ^{18}O depletions by several per mill (Lawrence 1989). The major reason for this ^{18}O depletion seems to be a low-temperature exchange with the oceanic crust in the underlying rock sequence. The ^{18}O shift in the solid phases is mostly due to an increase in temperature.

The situation for carbon is similar to meteoric diagenesis. ^{13}C-depleted carbon may originate from oxidation of organic matter, but at greater depths methane production can supply CO_2/HCO_3^- enriched in ^{13}C.

Isotopic data on several thousand limestone samples have been reported in the literature to date. The tendency toward lower $^{18}O/^{16}O$ ratios with increasing age is a well-documented fact (Keith and Weber 1964; Veizer and Hoefs 1976), although the reasons for this isotope shift are still under debate (see discussion on the isotopic evolution of ocean water in Sect. 3.7). Earlier limestone studies utilized whole-rock samples, but in recent years individual components, such as different generations of cements, have been analyzed (Hudson 1977; Dickson and Coleman 1980; Moldovany and Lohmann 1984; Given and Lohmann 1985; Dickson et al. 1990). These studies suggest that early cements exhibit higher $\delta^{18}O$- and $\delta^{13}C$-values, with successive cements becoming progressively depleted in both ^{13}C and ^{18}O. The ^{18}O trend may be due to increasing temperatures and to isotopic evolution of pore waters. Employing a laser ablation technique, Dickson et al. (1990) identified a very fine-scale O-isotope zonation in calcite cements, which they interpreted as indicating changes in the isotope composition of the pore fluids.

A more unusual effect of diagnesis is the formation of carbonate concretions in argillaceous sediments. Isotope studies by Hoefs (1970), Sass and Kolodny (1972), Irwin et al. (1977), and Gautier (1982) suggest that micobiological activity created localized supersaturation of calcite in which dissolved carbonate species were produced more rapidly than they could be dispersed by diffusion. Extremely variable $\delta^{13}C$-values in these concretions indicate that different microbiological processes participated in concretionary growth. Irwin et al. (1977) presented a model in which organic matter is diagenetically modified in sequence by (1) sulfate reduction, (2) fermentation, and (3) thermally induced abiotic CO_2 formation, which can be distinguished on the basis of their $\delta^{13}C$ values: (1) –25, (2) +15, and (3) –20‰.

3.11.6
Dolomites

The "dolomite problem" – meaning the origin of dolomite and the conditions promoting the dolomitization of limestones – is still being debated (Hardie 1987). Since dolomitization takes place in the presence of water, oxygen isotope compositions are determined by the pore fluid composition and by the temperature of formation. Carbon isotope compositions, in contrast, are determined by the precursor carbonate composition, because pore fluids generally have low carbon contents so that the $\delta^{13}C$-value of the precursor is retained. Two problems complicate the interpretation of isotope data to delineate the origin and diagenesis of dolomites. First, it has not been possible to determine the equilibrium oxygen isotope fractionations between dolomite and water at sedimentary temperatures directly, because the synthesis of dolomite at these low temperatures is problematic. Extrapolations of high-temperature experimental dolomite-water fractionations to low temperatures suggest that at 25 °C dolomite should be enriched in ^{18}O relative to calcite by 4–7‰ (e.g., Sheppard and Schwarcz 1970). By contrast, the oxygen isotope fractionation observed between Holocene calcite and dolomite is somewhat lower, namely in the range between 2‰ and 4‰ (Land 1980; McKenzie 1984). Secondly, the fractionation may depend partly on the crystal structure, more specifically on the composition and the degree of crystalline order; and, in this respect, dolomite is a very complex mineral.

Figure 68 summarizes oxygen and carbon isotope compositions of some recent and Pleistocene dolomite occurrences (after Tucker and Wright 1990). Variations in oxygen isotope composition reflect the involvement of different types of waters (from marine to fresh waters) and varying ranges of temperatures. With respect to carbon, $\delta^{13}C$-values between 0 and 3‰ are typical of marine signatures. In the presence of abundant organic matter, negative $\delta^{13}C$-values in excess of –20‰ indicate that carbon is derived from the decomposition of organic matter. Very positive $\delta^{13}C$-values up to +15‰ result from fermentation of organic matter (Kelts and McKenzie 1982). Such isotopically heavy dolomites have been described, for example, from the Guaymas Basin, where dolomite formation has taken place in the zone of active methanogenesis.

Fig. 68. Oxygen and carbon isotope compositions of some recent and Pleistocene dolomite occurrences. (After Tucker and Wright 1990)

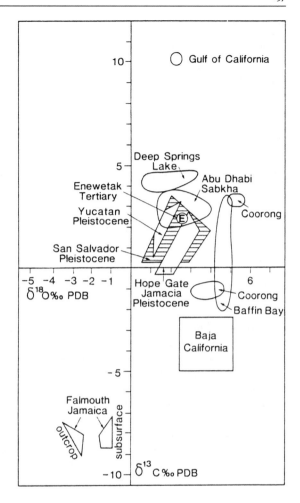

3.11.7
Freshwater Carbonates

Carbonates deposited in freshwater lakes exhibit a wide range in isotopic composition, depending upon the isotopic composition of the rainfall in the catchment area, its amount and seasonality, the temperature of evaporation, the relative humidity, and the biological productivity. Lake carbonates typically consist of a matrix of discrete components, such as detrital components, authigenic precipitates, and neritic and benthonic organisms. The separate analysis of such components has the potential to permit investigation of the entire water column, and the use of species with known seasonal life histories might make the study of seasonal variations possible. For example, the oxygen isotopic composition of authigenic carbonates can be used to obtain a surface water signal of changes in temperature and meteoric conditions, while the composition of bottom dwellers can be used

Fig. 69. Carbon and oxygen isotope compositions of freshwater carbonates from recent "closed" lakes. (After Talbot 1990; Reprinted from Chemical Geology, Vol. 80 No. 4, A review of the palaeohydrological interpretation of carbon and oxygen isotopic ratios in primary lacustrine carbonates, pp 261–279, © 1990, with kind permission from Elsevier Science – NL, Sara Burgerhartstraat 25, 1055 KV Amsterdam, The Netherlands)

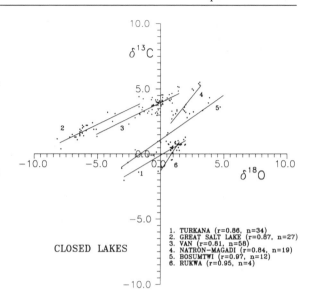

1. TURKANA (r=0.86, n=34)
2. GREAT SALT LAKE (r=0.87, n=27)
3. VAN (r=0.81, n=58)
4. NATRON–MAGADI (r=0.84, n=19)
5. BOSUMTWI (r=0.97, n=12)
6. RUKWA (r=0.95, n=4)

CLOSED LAKES

as a monitor of the water composition, assuming that the bottom water temperatures remained constant. To date, however, most studies have involved the analysis of only one or two forms of the lacustrine carbonate matrix.

The carbon and oxygen isotope compositions of carbonate precipitates from many lakes show a strong covariance with time, typically in those lakes which represent closed systems or water bodies with long residence times (Talbot 1990). In contrast, weak or no temporal covariance is typical of lakes which represent open systems with short residence times. Figure 69 gives examples of such covariant trends. Each closed lake appears to have a unique isotopic identity defined by its covariant trend, which depends on the geographical and climatic setting of a basin, its hydrology, and the history of the water body (Talbot 1990).

3.11.8
Phosphates

As was pointed out by Urey et al. (1951), the development of a temperature scale using $CaCO_3$ and another oxygen compound precipitated by marine organisms (e.g., phosphates) would permit a temperature scale independent of the oxygen isotopic composition of ocean water. However, subsequently shown by Longinelli and Nuti (1973) and others, the slope of the phosphate-water fractionation is practically identical to that of the carbonate-water fractionation. This means that the difference in the $^{18}O/^{16}O$ ratios of carbonate and phosphate is constant and, therefore, independent of the temperature and of the ^{18}O content of the water. Several studies have reported phosphate–water temperature relationships on the basis of various experimental results (Longinelli and Nuti 1973; Kolodny et al. 1983; Karhu and Epstein 1986; Shemesh et al. 1988), which fairly agree over the temper-

Fig. 70. Histogram of $\delta^{18}O$-values for cherts, limestones, and phosphorites of the Mishash Formation of Campanian age in Israel. *Arrows* mark expected composition of cherts and carbonates in equilibrium with phosphate of $\delta^{18}O=19.5‰$. (Shemesh et al. 1983)

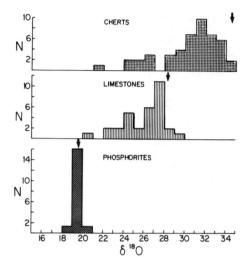

ature range from 10 to 60 °C. However, none of these relationships can be considered to represent a rigorously calibrated thermometer.

The major advantage of the phosphate thermometer is that it is a system which is not as sensitive toward diagenetic reactions as carbonates. This relative inertness towards post-depositional recrystallization is illustrated in Fig. 70, where the isotopic composition of cherts, limestones, and phosphorites from a single area in the Negev, Israel, is compared (Shemesh et al. 1983): the $\delta^{18}O$-values of carbonates and cherts vary widely, with probably only the highest values indicative of primary precipitation conditions. By contrast, the phosphorites are very uniform in isotope composition. The small spread in phosphorites is an argument for the resistance of the phosphate ion during diagenesis. Another interesting observation made by Shemesh et al. (1983) is that the $\delta^{18}O$-values of phosphorites decrease with increasing geological age, a similar trend to that observed for limestones and cherts (see Sect. 3.11.3 and 3.11.5).

The potential palaeoclimatic use of $\delta^{18}O$-values of mammal bone phosphate has been discussed by Longinelli (1984), Luz and Kolodny (1985), Ayliffe and Chivas (1990), besides others. These studies have demonstrated that $\delta^{18}O$ of bone phosphate is a monitor of the ^{18}O content of the body fluid, which in turn is determined by the O-isotope composition of drinking water.

3.11.9
Iron Oxides

The isotopic composition of most low-temperature minerals is generally well known. This is not so for the iron oxides/hydroxides, but due to the efforts of Yapp and coworkers this deficiency is beginning to change (Yapp 1983, 1987; Yapp and Pedley 1985). Yapp (1987) concluded that goethite forms in equilibrium with locally derived water. A plot of δD- versus $\delta^{18}O$-values of different minerals reveals that

Fig. 71. Plot of hydrogen and oxygen isotope compositions of goethite, kaolinite, and chert relative to meteoric waters. (After Yapp 1987; Reprinted from Geochimica et Cosmochimica Acta, Vol. 51 No. 2, Oxygen and hydrogen isotope variations among goethites (α=FeOOH) and the determination of paleotemperatures, pp 355-364, © 1987, with kind permission from Elsevier Science Ltd, The Boulevard, Langford Lane, Kidlington OX5 1GB, UK)

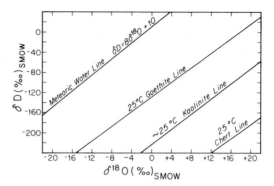

goethite exhibits the smallest fractionations relative to the Meteoric Water Line, while chert has the largest fractionation relative to water (Fig. 71). Yapp (1987) postulated that goethite is retentive of its orginal isotope composition through time and thus can be used as a geothermometer in combination with other low-temperature minerals.

During conversion to hematite only small fractionation effects seem to occur, because most of the oxygen remains in the solid (Yapp 1987). Thus, in principle it should be possible to reconstruct the sedimentary environment of iron oxides from Precambrian Banded Iron Formations (BIFs). By analyzing the least metamorphosed BIFs, Hoefs (1992) concluded that the situation is not so simple. Infiltration of external fluids during diagenesis and/or low-temperature metamorphism appears to have erased the primary isotope record in these ancient sediments.

3.11.10
Sedimentary Sulfur

Analysis of the sulfur isotope composition of sediments may yield important information about the origin and diagenesis of sulfur compounds. Due to the activity of sulfate-reducing bacteria, most sulfur isotope fractionation takes place in the uppermost mud layers in shallow seas and tidal flats. As a result, sedimentary sulfides are depleted in ^{34}S relative to ocean water sulfate. The depletion is usually in the order of 20‰–60‰ (Hartmann and Nielsen 1969; Goldhaber and Kaplan 1974), although bacteria in pure cultures have been observed to produce fractionations of only 10‰–30‰, with a maximum reported value of 46‰ (Kaplan and Rittenberg 1964).

Bacterial sulfate reduction is accomplished by the oxidation of organic matter:

$$2CH_2O + SO_4^{2-} \rightarrow H_2S + 2HCO_3^-$$

the resulting H_2S reacting with available iron, which is in the reactive nonsilicate bound form (oxy-hydroxides). Thus, the amount of pyrite formed in sediments may be limited by (1) the amount of sulfate, (2) the amount of organic matter, and (3) the amount of reactive iron. Based upon the relationships between these three reservoirs, different scenarios for pyrite formation in anoxic environments can be envisaged (Raiswell and Berner 1985).

Early workers have shown that pyrite is the main sedimentary sink for reduced sulfur, but the rate of H_2S production in many sediments exceeds the rate at which it can be removed by reaction with iron minerals (Chanton et al. 1987a,b). In such cases H_2S undergoes further reactions; some H_2S diffuses back into surface sediments where it is reoxidized, and some H_2S may be incorporated into sedimentary organic matter. Chanton et al. (1987b) demonstrated that small variations in the inward flux of sulfate and the outward flux of sulfide can dramatically change the isotopic composition of total buried sulfide.

In recent years there has been much progress to identify and measure the isotopic composition of different forms of sulfur in sediments (e.g., Mossmann et al. 1991; Zaback and Pratt 1992). Pyrite is generally considered to be the end product of sulfur diagenesis in anoxic marine sediments. Acid-volatile sulfides (AVS), which include "amorphous" FeS, mackinawite, greigite, and pyrrhotite, are considered to be a transient early species, but recent investigations by Mossmann et al. (1991) have demonstrated that AVS can form before, during, and after precipitation of pyrite within the upper tens of centimeters of sediment.

Up to six sulfur species (disulfide, acid-volatile sulfide, kerogen, bitumen, sulfate, and elemental sulfur) have been separated and isotopically analyzed by Zaback and Pratt (1992). Their data provide information regarding the relative timing of sulfur incorporation and the sources of the individual sulfur species. Pyrite exhibits the greatest [34]S depletion relative to seawater. Acid-volatile sulfur and kerogen sulfur are generally enriched in [34]S relative to pyrite. This indicates that pyrite precipitated nearest to the sediment-water interface under mildly reducing conditions, while AVS and kerogen sulfur resulted from formation at greater depth under more reducing conditions with low concentrations of pore water sulfate. Elemental sulfur is most abundant in surface sediments and, probably, formed by oxidation of sulfide diffusing across the sediment-water interface.

Besides bacterial sulfate reduction, thermochemical sulfate reduction in the presence of organic matter is another process which can produce large quantities of H_2S. The crucial question is whether abiological sulfate reduction can occur at temperatures as low as 100 °C, which is just above the limit of microbiological reduction. Trudinger et al. (1985) concluded that abiological reduction below 200 °C had not been unequivocally demonstrated, although they did not dismiss its possible significance. In recent years, the evidence for thermochemical sulfate reduction, even at temperatures near 100 °C or lower, has increased (e.g., Krouse et al. 1988; Machel et al. 1995). The characteristic feature of thermochemical sulfate reduction is very small sulfur isotope fractionation relative to sulfate. Thus, it is likely that this process is much more prevalent than originally thought.

3.12
Metamorphic Rocks

The isotope composition of metamorphic rocks is mainly controlled by three factors besides the temperature of exchange. These are (1) the composition of the pre-metamorphic protolith, (2) the effects of volatilization with increasing temperatures, and (3) an exchange with infiltrating fluids or melts. The relative importance of these three factors can vary extremely from area to area and from rock type to rock type; and the accurate interpretation of the causes of isotope variations in metamorphic rocks requires knowledge of the reaction history of the respective metamorphic rocks.

1. The isotope composition of the precursor rock – either sedimentary or magmatic – is usually very difficult to estimate. Only in very dry environments do metamorphic rocks retain their original signatures.

2. Prograde metamorphism of sediments causes the liberation of volatiles, which can be described by two end-member processes (Valley 1986): (1) Batch volatilization, where all fluid is evolved before any is permitted to escape, and (2) Rayleigh volatilization, which requires that once fluid is generated it is isolated immediately from the rock. Natural processes seem to fall between these extreme situations; nevertheless these two end-member situations provide useful limits. Metamorphic volatilization reactions generally reduce the $\delta^{18}O$-value of a rock because CO_2 and, in most cases, H_2O lost are enriched in ^{18}O. The magnitude of ^{18}O depletion can be estimated by considering the relevant fractionations at the respective temperatures. In any case the effect on the $\delta^{18}O$-value should be small (around 1‰), because the amount of oxygen liberated is small compared to the remaining oxygen in the rock and fractionations at these rather high temperatures are small and, in some cases, may even reverse sign.

3. The infiltration of externally derived fluids is a controversal idea, but has gained much support in recent years. Many studies have convincingly demonstrated that a fluid phase plays a far more active role than was previously envisaged (e.g., Ferry 1992).

Fluid flow normally occurs along temperature gradients. The isotopic composition of both rock and fluid must continually change in instances where temperature increases or decreases along the flow path and where local isotopic exchange equilibrium is approached. Therefore, fluid flow along temperature gradients is potentially a mechanism for explaining stable isotope variations in many metamorphic terranes (e.g., Dipple and Ferry 1992).

A critical issue is the extent to which the isotope composition of a metamorphic rock is modified by a fluid phase. Volatilization reactions leave an isotope signature greatly different from that produced when fluid-rock interaction accompanies mineral-fluid reaction. Changes of 5‰–10‰ are a strong indication that fluid-rock interaction rather than volatilization reactions occurred during the metamorphic event. Two end-member situations can be postulated in which coexisting minerals would change their isotopic composition during fluid-rock interaction:

1. A pervasive fluid moves independently of structural and lithologic control through a rock and leads to a homogenization of whatever differences in isotopic composition may have existed prior to metamorphism.

2. A channelized fluid leads to local equilibration on the scale of individual beds or units, but does not result in isotopic homogenization of all rocks or units. Channelized flow favors chemical heterogeneity, allowing some rocks to remain unaffected. Although both types of fluid flow appear to be manifest in nature, the latter type appears to be more common.

The behavior of volatiles with increasing temperatures during prograde metamorphism can also be studied by analyzing nitrogen isotope compositions. Data for progressively metamorphosed sedimentary rocks have documented a decrease in N concentrations and an increase in ^{15}N contents (Haendel et al. 1986; Bebout and Fogel 1992). These systematic trends can be explained by a devolatilization process approximating Rayleigh distillation and an N_2-NH_4^+ exchange as the dominant mechanisms of N-isotope fractionation. Thus, N isotopes may provide a valuable parameter of evaluating open- and closed-system behavior during metamorphic devolatilization and may as well be an effective tracer of large-scale volatile transport.

Also, sulfur isotope compositions may potentially be modified during regional metamorphism via a mobile phase. However, as shown by Oliver et al. (1992), sulfur isotope data from regionally metamorphosed graphitic sulfidic schists on average have $\delta^{34}S$-values of -27‰, suggesting that the rocks have not been disturbed from their inferred organic-rich sedimentary origin. Only at temperatures >500 °C do matrix pyrite-pyrrhotite pairs approach isotopic equilibrium at millimeter to centimeter scales, suggesting that the process that favored equilibrium was recrystallization.

Two of the very exciting recent trends in the analysis of metamorphic rocks have been technological advances in microanalytical techniques and theoretical advances, such as the modeling of isotope discontinuities (fronts) and the numerical modeling of isotope exchange among minerals.

Taylor and Bucher-Nurminen (1986), for instance, report sharp isotopic gradients (fronts) of up to 17‰ in $\delta^{18}O$ and 7‰ in $\delta^{13}C$ over distances of a few millimeters in calcite around veins in the contact aureole of the Bergell granite. Similar sharp gradients have also been observed in other metasomatic zones but are often unrecognized because an unusually detailed millimeter-scale sampling is required.

Well-defined stable isotope profiles may be used to provide quantitative information on fluid fluxes such as the direction of fluid flow and the duration of infiltration events (Baumgartner and Rumble 1988; Ganor et al. 1989; Bickle and Baker 1990; Cartwright and Valley 1991; Dipple and Ferry 1992). In well-constrained situations, fluid flow modeling permits estimation of fluid fluxes that are far more realistic than fluid/rock ratios calculated from a zero-dimensional model. Estimates of integrated fluid fluxes range from less than $0.1 \text{ m}^3/\text{m}^2$ to as much as $10^5 \text{ m}^3/\text{m}^2$ (Ganor et al. 1989; Ferry 1992), which must reflect real differences in permeabilities between metamorphic rock units.

Due to the invention of new microanalytical techniques (laser sampling and ion microprobe), it has become possible to document small-scale isotope gradients within single mineral grains. Oxygen isotope zoning in garnet (Chamberlain and Conrad 1991; Jamtveit and Hervig 1994) and in magnetite (Eiler et al. 1995) are especially pronounced. Changes of 3–5‰ in $\delta^{18}O$-value from core to rim may occur during garnet growth as a result of the infiltration of fluids. From ion microprobe analyses, Eiler et al. (1995) presented evidence that closed system diffusion may produce internal zonation in magnetite. The shape of the isotopic gradient across a grain contact will allow distinction among processes controlled by open-system fluid migration or closed-system diffusion, and may also help to interpret inconsistencies in geothermometric data.

3.12.1
Regional Metamorphism

It is a general observation that low-grade metamorphic pelites have $\delta^{18}O$-values between 15 and 18‰, whereas high-grade gneisses have $\delta^{18}O$-values between 6 and 10‰ (Garlick and Epstein 1967; Shieh and Schwarcz 1974; Longstaffe and Schwarcz 1977 and others). This general tendency for a decrease in ^{18}O on the order of 5‰ and more with increasing metamorphic grade is a matter of intense debate (Rye et al. 1976; Wickham and Taylor 1985; Peters and Wickham 1995). In the absence of infiltration of a fluid phase, isotopic shifts resulting from net transfer reactions in typical amphibolite or lower granulite facies metapelites and metabasites are about 1‰ or less for about 150 °C of heating (Kohn 1993). Thus, the processes responsible for this decrease in ^{18}O must be linked to large-scale fluid transport in the crust .

There are several factors which control fluid transport. One is the lithology of a metamorphic sequence. Marbles, in particular, are relatively impermeable during metamorphism (Nabelek et al. 1984) and, therefore, may act as barriers to fluid flow, limiting the scale of homogenization and preferentially channeling fluids through silicate layers. Marbles may act as local high–^{18}O reservoirs and may even increase the ^{18}O content of adjacent lithologies (Peters and Wickham 1995). Therefore, massive marbles generally preserve their sedimentary signatures, even up to the highest metamorphic grades (Valley et al. 1990). Controversy remains about fluid sources and large-scale transport mechanisms during regional metamorphism.

Sedimentary sequences undergoing a first metamorphic event initially may contain abundant connate pore fluids which provide a substantial low-^{18}O reservoir and a medium for isotopic homogenization. An additional important fluid source is metamorphic dehydration reactions (e.g., Ferry 1992). In some areas, petrological and stable isotope studies suggest that metamorphic fluid compositions were predominantly internally buffered by devolatilization reactions and that large amounts of fluid did not interact with the rocks during regional metamorphism (e.g., Valley et al. 1990). In a polymetamorphic terrane, later metamorphic events are likely to be dominated by magmatic fluid sources since previous events would have caused extensive dehydration, thereby limiting potential fluid sources (Peters and Wickham 1995). A detailed study of the O-isotope composition of pelites, amphibolites, and marbles from the island of Naxos, Greece, exhibits that the isotopic pattern observed today is the result of at least three processes: two fluid flow events and a pre-existing isotopic gradient (Baker and Matthews 1995).

Shear zones are particularly good environments to investigate fluid flow at various depths within the crust (Kerrich et al. 1984; Kerrich and Rehrig 1987; McCaig et al. 1990; Fricke et al. 1992). During retrograde metamorphism aqueous fluids react with dehydrated rocks and fluid flow is concentrated within relatively narrow zones. By analyzing quartzite mylonites in Nevada, Fricke et al. (1992) demonstrated that significant amounts of meteoric waters must have infiltrated the shear zone during mylonitization to depths of at least 5–10 km. Similarly, McCaig et al. (1990) showed that formation waters were involved in shear zones in the Pyrenees and that the mylonitization process occurred at a depth of about 10 km.

3.12.2
Lower Crustal Rocks

Granulites constitute the dominant rock type in the lower crust and may be found at the Earth's surface in two different settings: (1) exposed in high-grade regional metamorphic belts and (2) found as small xenoliths in basaltic pipes. Both types of granulites suggest a compositionally diverse lower crust ranging in composition from mafic to felsic.

Stable isotope studies of granulite terranes (Sri Lanka – Fiorentini et al. 1990, South India – Jiang et al. 1988, Limpopo Belt – Hoernes and Van Reenen 1992; Venneman and Smith 1992, Adirondacks – Valley and coworkers) have shown that terranes are isotopically heterogeneous and are characterized by $\delta^{18}O$-values that range from "mantle-like" values to typical metasedimentary values above 10‰. Investigations of amphibolite/granulite transitions have shown little evidence for a pervasive fluid flux as a major factor in granulite facies metamorphism (Valley et al. 1990; Baker 1990; Cartwright and Valley 1991; Todd and Evans 1993).

Similar results have been obtained from lower crustal granulite xenoliths, which also exhibit a large range in $\delta^{18}O$-values from 5.4 to 13.5‰ (Fowler and Harmon 1990; Mengel and Hoefs 1990; Kempton and Harmon 1992). Mafic granulites are characterized by the lowest $\delta^{18}O$-values and range of ^{18}O contents. By contrast, silicic meta-igneous and meta-sedimentary granulites are significantly enriched in ^{18}O, with an average $\delta^{18}O$-value around 10‰. The overall variation of 8‰ emphasizes the O-isotope heterogeneity of the lower crust and demonstrates that pervasive deep crustal fluid flow and isotopic homogenization is not a major process.

There has been considerable debate on the source and role of CO_2 in granulite genesis since the discovery of high-density CO_2 in high-grade metamorphic rocks by Touret (1971). Information on the source of the CO_2 may be obtained from carbon isotope analysis. Several studies of this kind have been undertaken, with the

Fig. 72. Histogram of C-isotope composition of fluid inclusion and chemically bound CO_2 from different granulite terrains. (After Iyer et al. 1995)

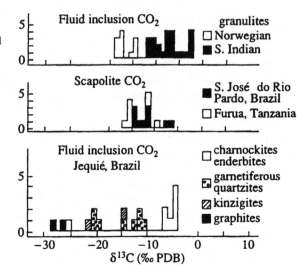

results being summarized in Fig. 72. As indicated in Fig. 72, δ^{13}C-values span a large range from $-25‰$ to about $0‰$. Such a large range is not compatible with a unique carbon source, but favors the participation of different sources of CO_2 in granulite formation.

3.12.3
Contact Metamorphism

Because the isotopic composition of igneous rocks is quite different from that of sedimentary rocks, studies of the isotope variations in the vicinity of an intrusive contact offer the opportunity to investigate the role of fluids interacting with rocks around cooling plutons. Two types of aureoles can be distinguished (Nabelek 1991): (1) "closed" aureoles where fluids are derived from the pluton or the wall-rock and (2) "open" aureoles which for at least part of their metamorphic history are infiltrated by fluids of an external derivation. Some aureoles will be dominated by magmatic or metamorphic fluids, whereas others are dominated by surface-derived fluids. The occurrence of meteoric-hydrothermal systems around many plutonic complexes has been described by H.P. Taylor and his coworkers and has been described in more detail on p. 95. The depth to which surface-derived fluids can penetrate is still under debate, but most meteoric-hydrothermal systems appear to have occurred at depths less than ≈ 6 km (Criss and Taylor 1986). However, Wickham and Taylor (1985) suggested that seawater infiltration has been observed to a depth of 12 km in the Trois Seigneur Massif, Pyrenees.

In many contact aureoles combined petrologic and isotope studies have provided evidence that fluids were primarily locally derived. Oxygen isotope composition of calc-silicates from many contact aureoles has revealed that the ^{18}O contents of the calc-silicate hornfelses approach those of the respective intrusions. This, together with characteristic hydrogen and carbon isotope ratios, has led many workers to conclude that magmatic fluids were dominant during contact metamorphism (Taylor and O'Neil 1977; Nabelek et al. 1984; Bowman et al. 1985; Valley 1986). Ferry and Dipple (1992) developed different models to simulate fluid-rock interaction on the Notch Peak aureole, Utah. Their preferred model assumes fluid flow in the direction of increasing temperature and thus argues against magmatic fluids, but instead for fluids derived from volatilization reactions. Nabelek (1991) calculated model δ^{18}O-profiles which should result from both "down-temperature" and "up-temperature" flow in a contact aureole. He demonstrated that the presence of complex isotopic profiles can be used to obtain information about fluid fluxes. Gerdes et al. (1995) have examined meter-scale ^{13}C and ^{18}O transport in a thin marble layer near a dike in the Adamello contact aureole, southern Alps, and observed systematic stable isotope changes in the marble over <1 m as the dike is approached, with δ^{13}C-values ranging from 0 to $-7‰$ and δ^{18}O-values from 22.5 to 12.5‰, respectively. These authors have compared the isotope profiles to one- and two-dimensional models of advective-dispersive isotope transport. Best agreement is obtained using a two-dimensional model that specifies (1) a high permeability zone flow and (2) a lower permeability zone in marble away from the dike.

3.12.4
Thermometry

Oxygen isotope thermometry is widely used to determine temperatures of meta-morphic rocks. The principal concern in isotope thermometry continues to be the evidence of preservation of peak metamorphic temperatures during cooling. It has long been recognized that oxygen isotope thermometers often record dis-cordant temperatures in slowly cooled metamorphic rocks. Giletti (1986) pointed out that, in a rock consisting of three or more minerals with different oxygen clo-sure temperatures, most mineral pairs record disequilibrium temperatures. These apparent temperatures may be higher or lower than the maximum temperature experienced by the rock.

The resetting of isotope thermometers results from diffusional re-equilibra-tion during cooling (Deines 1977; Giletti 1986; Valley 1986; Eiler et al. 1992, 1993). Assuming that a rock behaves as a closed system and consists of the three miner-al assemblage quartz, feldspar, and horblende, then hornblende will be the slow-est diffusing phase and feldspar the fastest diffusing phase. Using the formulation of Dodson (1973) for closure temperature and a given set of parameters (diffusion constants, cooling rate and grain size), Giletti (1986) calculated apparent temper-atures that would be obtained in rocks with different modal proportions of the three minerals once all isotope exchange had ceased in the rock. In the Giletti model, the apparent quartz–hornblende temperature is dependent only on the quartz/feldspar ratio and is independent of the amount of hornblende in the rock. Recently, Eiler et al. (1992, 1993) demonstrated that the abundance of the slow-diffusing phase (e.g., hornblende) has a large effect on apparent tempera-tures because of continued exchange between the grain boundaries of this phase and fast-diffusing phases.

In conclusion, best samples for isotope thermometry are those that have at least one slow-diffusing phase and a high modal proportion of the fast-diffusing phase. The maximum temperatures recovered from such samples are limited on-ly by the closure temperature of the high-closing phase. Garnet and spinel appear to be the best candidates from which isotope temperature data over 950 °C can be retrieved (Farquhar et al. 1993). A drawback for garnet is the possibility that pro-grade isotopic zoning will be preserved, thus complicating temperature interpre-tations. Pyroxenes and iron oxides are not as good as garnet or spinel because they have higher oxygen diffusivities and are more susceptible to late alteration. Farquhar et al. (1996) have investigated two granulite terrains from northwest Canada and Antarctica. Quartz-garnet temperatures of around 1000 °C are in good agreement with a variety of independent temperature estimations. Quartz-pyroxene temperatures are significantly lower and still lower quartz-magnetite temperatures of around 670 °C are attributed to a combination of faster oxygen diffusion in quartz and magnetite and recrystallization during late-stage defor-mation. The "dry" nature of granulites is obviously critical for preservation of high-temperature records. Cooler and more hydrous rocks seem be less capable of retaining a record of peak temperatures.

Other suitable phases for the preservation of peak metamorphic temperatures are the Al_2SiO_5 polymorphs kyanite and sillimanite, both having slow oxygen dif-

Fig. 73. Temperature dependence of calcite-graphite fractionations as determined by various methods. (After Kitchen and Valley 1995)

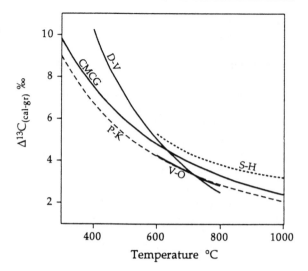

fusion rates. By analyzing the alumosilicate polymorphs from a variety of rocks with different temperature histories, Sharp (1995) was able to derive equilibrium fractionation factors for kyanite and sillimanite. In some rocks oxygen isotope temperatures are far higher than the regional metamorphic temperatures, possibly reflecting early high-temperature contact metamorphic effects that are preserved only in the most refractory phases.

Carbon isotope partitioning between calcite and graphite is another example of a favorable thermometer to record peak metamorphic temperatures in marbles because calcite is the abundant phase with relatively high carbon diffusivities whereas graphite is of minor abundance and has a very slow diffusion rate. The calcite–graphite pair has been calibrated by various authors and various approaches (Fig. 73). Although a fairly good agreement among the various calibration curves exists in the temperature range from 600 to 800 °C, the remaining inconsistencies cannot be solved at present (Kitchen and Valley 1995). Figure 73 demonstrates the inconsistencies of even recently published temperature curves derived from empirical, theoretical, and experimental calibrations. Nevertheless there is hope that with more and more sophisticated calibration and sampling techniques it will soon be possible to establish "true" metamorphic peak temperatures.

References

Abelson PH, Hoering TC (1961) Carbon isotope fractionation in formation of amino acids by photosynthetic organisms. Proc Natl Acad Sci USA 47:623

Abrajano TA, Sturchio NB, Bohlke JH, Lyon GJ, Poreda RJ, Stevens MJ (1988) Methane – hydrogen gas seeps Zambales ophiolite, Phillippines: deep or shallow origin. Chem Geol 71:211–222

Agyei EK, McMullen CC (1968) A study of the isotopic abundance of boron from various sources. Can J Earth Sci 5:921–927

Allard P (1983) The origin of hydrogen, carbon, sulphur, nitrogen and rare gases in volcanic exhalations: evidence from isotope geochemistry. In: Tazieff H, Sabroux JC (eds) Forecasting volcanic events. Elsevier, Amsterdam, pp 337–386

Altabet MA (1988) Variations in nitrogen isotopic compositions among particle classes: implications for particle transformation and flux in the open ocean. Deep-Sea Res 35:535–544

Altabet MA, Deuser WC (1985) Seasonal variations in natural abundance of ^{15}N in particles sinking to the deep Sargasso Sea. Nature 315:218–219

Altabet MA, McCarthy JJ (1985) Temporal and spatial variations in the natural abundance of ^{15}N in POM from a warm-core ring. Deep Sea Res 32:755–772

Altabet MA, Deuser WG, Honjo S, Stienen C (1991) Seasonal and depth related changes in the source of sinking particles in the North Atlantic. Nature 354:136–139

Amari S, Hoppe P, Zinner E, Lewis RS (1993) The isotopic compositions of stellar sources of meteoritic graphite grains. Nature 365:806–809

Amberger A, Schmidt HL (1987) Natürliche Isotopengehalte von Nitrat als Indikatoren für dessen Herkunft. Geochim Cosmochim Acta 51:2699–2705

Anderson AT, Clayton RN, Mayeda TK (1971) Oxygen isotope thermometry of mafic igneous rocks. J Geol 79:715–729

Anderson TF, Arthur MA (1983) Stable isotopes of oxygen and carbon and their application to sedimentologic and paleoenvironmental problems. In: Stable isotopes in sedimentary geology. SEPM short course no 10, Dallas, 1983, pp 111–151

Arnold M, Sheppard SMF (1981) East Pacific Rise at 21oN: isotopic composition and origin of the hydrothermal sulfur. Earth Planet Sci Lett 56:148–156

Arthur MA, Dean WE, Claypool CE (1985) Anomalous ^{13}C enrichment in modern marine organic carbon. Nature 315:216–218

Ayliffe LK, Chivas AR (1990) Oxygen isotope composition of the bone phosphate of Australian kangaroos: potential as a palaeoenvironmental recorder. Geochim Cosmochim Acta 54:2603–2609

Bachinski DJ (1969) Bond strength; and sulfur isotope fractionation in coexisting sulfides. Econ Geol 64:56–65

Baertschi P (1976) Absolute ^{18}O content of standard mean ocean water. Earth Planet Sci Lett 31:341–344

Bainbridge KT, Nier AO (1950) Relative isotopic abundances of the elements. Preliminary Rep No 9. Nuclear Sciences Series, Natl Res Council USA, Washington DC

Bains-Sahota SK, Thiemens MH (1989) A mass independent sulfur isotope effect in the nonthermal formation of S_2F_{10}. J Chem Phys 90:6099–6110

Baker AJ (1990) Stable isotope evidence for fluid-rock interactions in the Ivrea Zone, Italy. J Petrol 31:243–260

Baker AJ, Fallick AE (1989) Heavy carbon in two-billion-year-old marbles from Lofoten-Vesteralen, Norway: implications for the Precambrian carbon cycle. Geochim Cosmochim Acta 53:1111–1115

Baker J, Matthews A (1995) The stable isotopic evolution of a metamorphic complex, Naxos, Greece. Contr Mineral Petrol 120:391–403

Banner JL, Wasserburg GJ, Dobson PF, Carpenter AB, Moore CH (1989) Isotopic and trace element constraints on the orgin and evolution of saline groundwaters from central Missouri. Geochim Cosmochim Acta 53:383–398

Barnes I, Irwin WP, White DE (1978) Global distribution of carbon dioxide discharges and major zones of seismicity. US Geol Survey, Water-Resources Investigation 78–39, Open File Report

Barth S (1993) Boron isotope variations in nature: a synthesis. Geol Rundsch 82:640–651

Bassett RL (1990) A critical evaluation of available measurements for the stable isotopes of boron. Appl Geochem 5:541–554

Baumgartner LP, Rumble D (1988) Transport of stable isotopes. I. Development of a kinetic continuum theory for stable isotope transport. Contr Mineral Petrol 98:417–430

Beaty DW, Taylor HP (1982) Some petrologic and oxygen isotopic relationships in the Amulet Mine, Noranda, Quebec, and their bearing on the origin of Archaean massive sulfide deposits. Econ Geol 77:95–108

Beaudoin G, Taylor BE, Rumble D, Thiemens M (1994) Variations in the sulfur isotope composition of troilite from the Canyon Diablo iron meteorite. Geochim Cosmochim Acta 58:4253–4255

Bebout GE, Fogel ML (1992) Nitrogen isotope compositions of metasedimentary rocks in the Catalina Schist, California: implications for metamorphic devolatilization history. Geochim Cosmochim Acta 56:2839–2849

Becker RH, Epstein S (1982) Carbon, hydrogen and nitrogen isotopes in solvent-extractable organic matter from carbonaceous chondrites. Geochim Cosmochim Acta 46:97–103

Bender M L (1990) The $\delta^{18}O$ of dissolved O_2 in seawater: a unique tracer of circulation and respiration in the deep sea. J Geophys Res 95:22243–22252

Bender ML, Keigwin LD (1979) Speculations about upper Miocene changes in abyssal Pacific dissolved bicarbonate $\delta^{13}C$. Earth Planet Sci Lett 45:383–393

Berner RA (1990) Atmospheric carbon dioxide levels over Phanerozoic time. Science 249:1382–1386

Berner U, Faber E, Scheeder G, Panten D (1995) Primary cracking of algal and landplant kerogens: kinetic models of isotope variations in methane, ethane and propane. Chem Geol 126:233–245

Beveridge NAS, Shackleton NJ (1994) Carbon isotopes in recent planktonic foraminifera: a record of anthropogenic CO_2 invasion of the surface ocean. Earth Planet Sci Lett 126:259–273

Bhattacharya SK, Thiemens MH (1989) New evidence for symmetry dependent isotope effects: O+CO reaction. Z Naturforsch 44a:435–444

Bickle MJ, Baker J (1990) Migration of reaction and isotopic fronts in infiltration zones: assessments of fluid flux in metamorphic terrains. Earth Planet Sci Lett 98:1–13

Bickle MJ, Wickham SM, Chapman HJ, Taylor HP (1988) A strontium, neodymium and oxygen isotope study of hydrothermal metamorphism and crustal anatexis in the Trois Seigneurs massif, Pyrenees, France. Contr Mineral Petrol 100:399–417

Biemann K, Owen T, Rushneck DR, Lafleur AL, Howarth DW (1976) The atmosphere of Mars near the surface: isotope ratios and upper limits on noble gases. Science 194:76–78

Bigeleisen J (1965) Chemistry of isotopes. Science 147:463–471

Bigeleisen J, Mayer MG (1947) Calculation of equilibrium constants for isotopic exchange reactions. J Chem Phys 15:261–267

Bigeleisen J, Wolfsberg M (1958) Theoretical and experimental aspects of isotope effects in chemical kinetics. Adv Chem Phys 1:15–76

Bigeleisen J, Perlman ML, Prosser HC (1952) Conversion of hydrogenic materials for isotopic analysis. Anal Chem 24:1356

Bird MI, Chivas AR (1989) Stable-isotope geochronology of the Australian regolith. Geochim Cosmochim Acta 53:3239–3256

Bird MI, Chivas AR, Andrew AS (1989) A stable isotope study of lateritic bauxites. Geochim Cosmochim Acta 53:1411–1420

Bird MI, Longstaffe FJ, Fyfe WS, Bildgen P (1992) Oxygen isotope systematics in a multiphase weathering system in Haiti. Geochim Cosmochim Acta 56:2831–2838

Blattner P, Hulston JR (1978) Proportional variations of geochemical $\delta^{18}O$ scales – an interlaboratory comparison. Geochim Cosmochim Acta 42:59–62

Blattner P, Lassey KR (1989) Stable isotope exchange fronts, Damköhler numbers and fluid to rock ratios. Chem Geol 78:381–392

Blattner P, Dietrich V, Gansser A (1983) Contrasting ^{18}O enrichment and origins of High Himalayan and Transhimalayan intrusives. Earth Planet Sci Lett 65: 276–286

Boigk H, Hagemann HW, Stahl W, Wollanke G (1976) Isotopenphysikalische Untersuchungen zur Herkunft und Migration des Stickstoffs nordwestdeutscher Erdgase aus Oberkarbon und Rotliegend. Erdöl und Kohle 29:103–112

Borthwick J, Harmon RS (1982) A note regarding ClF_3 as an alternative to BrF_5 for oxygen isotope analysis. Geochim Cosmochim Acta 46:1665–1668

Bottinga Y (1969a) Calculated fractionation factors for carbon and hydrogen isotope exchange in the system calcite-carbon dioxide-graphite-methane-hydrogen-water-vapor. Geochim Cosmochim Acta 33:49–64

Bottinga Y (1969b) Carbon isotope fractionation between graphite, diamond and carbon dioxide. Earth Planet Sci Lett 5:301–307

Bottinga Y, Craig H (1969) Oxygen isotope fractionation between CO_2 and water and the isotopic composition of marine atmospheric CO_2. Earth Planet Sci Lett 5:285–295

Bottinga Y, Javoy M (1973) Comments on oxygen isotope geothermometry. Earth Planet Sci Lett 20:250–265

Bottinga Y, Javoy M (1975) Oxygen isotope partitioning among the minerals in igneous and metamorphic rocks. Rev Geophys Space Phys 13:401–418

Boudreau BP, Westrich JT (1984) The dependence of bacterial sulfate reduction on sulfate concentration in marine sediments. Geochim Cosmochim Acta 48:2503–2516

Bowers TS (1989) Stable isotope signatures of water-rock interaction in mid-ocean ridge hydrothermal systems: sulfur, oxygen, and hydrogen. J Geophys Res 94:5775–5786

Bowers TS, Taylor HP (1985) An integrated chemical and isotope model of the origin of mid-ocean ridge hot spring systems. J Geophys Res 90:12583–12606

Bowman JR, O'Neil JR, Essene EJ (1985) Contact skarn formation at Elkhorn, Montana. II. Origin and evolution of C-O-H skarn fluids. Am J Sci 285:621–660

Boyd AW, Brown F, Lounsbury M (1955) Mass spectrometric study of natural and neutron-irradiated chlorine. Can J Phys 33:35

Boyd SR, Pillinger CT (1994) A preliminary study of $^{15}N/^{14}N$ in octahedral growth from diamonds. Chem Geol 116:43–59

Boyd SR, Pillinger CT, Milledge HJ, Mendelssohn MJ, Seal M (1992) C and N isotopic composition and the infrared absorption spectra of coated diamonds: evidence for the regional uniformity of CO_2-H_2O rich fluids in lithospheric mantle. Earth Planet Sci Lett 109:633–644

Boyd SR, Hall A, Pillinger CT (1993) The measurement of $\delta^{15}N$ in crustal rocks by static vacuum mass spectrometry: application to the origin of the ammonium in the Cornubian batholith, southwest England. Geochim Cosmochim Acta 57:1339–1347

Brand W (1996) Isotope ratio monitoring techniques in mass spectrometry. J Mass Spectrometry (in press)

Bremner JM, Keeney DR (1966) Determination and isotope ratio analysis of different forms of nitrogen in soils, III. Soil Sci Soc Am Proc 30:577–582

Broecker WS (1974) Chemical oceanography. Harcourt Brace Jovanovich, New York

Broecker WS, Van Donk J (1970) Insolation changes, ice volumes and the ^{18}O record in deep-sea cores. Rev Geophys Space Phys 8:169–197

Brown FS, Baedecker MJ, Nissenbaum A, Kaplan IR (1972) Early diagenesis in a reducing fjord, Saanich Inlet, British Columbia. III. Changes in organic constituents of a sediment. Geochim Cosmochim Acta 36:1185–1203

Brumsack HJ, Zuleger E (1992) Boron and boron isotopes in pore waters from ODP Leg 127, Sea of Japan. Earth Planet Sci Lett 113:427–433

Brumsack HJ, Zuleger E, Gohn E, Murray RW (1992) Stable and radiogenic isotopes in pore waters from Leg 1217, Japan Sea. Proc Ocean Drilling Prog 127/128:635–649

Cameron EM (1982) Sulphate and sulphate reduction in early Precambrian oceans. Nature 296:145–148

Cameron EM, Hall GEM, Veizer J, Krouse HR (1995) Isotopic and elemental hydrogeochemistry of a major river system: Fraser River, British Columbia, Canada. Chem Geol 122:149–169

Canfield DE, Thamdrup B (1994) The production of ^{34}S depleted sulfide during bacterial disproportion to elemental sulfur. Science 266:1973–1975

Cartwright I, Valley JW (1991) Steep oxygen isotope gradients at marble-metagranite contacts in the NW Adirondacks Mountains, NY. Earth Planet Sci Lett 107:148–163

Cerling TE (1991) Carbon dioxide in the atmosphere: evidence from Cenozoic and Mesozoic paleosols. Am J Sci 291:377–400

Cerling TE, Brown FH, Bowman JR (1985) Low-temperature alteration of volcanic glass: hydration, Na, K, ^{18}O and Ar mobility. Chem Geol 52:281–293

Chamberlain CP, Conrad ME (1991) Oxygen isotope zoning in garnet. Science 254:403–404

Chambers LA, Trudinger PA (1979) Microbiological fractionation of stable sulfur isotopes. Geomicrobiology J 1:249–293

Chan LH (1987) Lithium isotope analysis by thermal ionization mass spectrometry of lithium tetraborate. Anal Chem 59:151–160

Chan LH, Edmond JM (1988) Variation of lithium isotope composition in the marine environment: a preliminary report. Geochim Cosmochim Acta 52:1771–1717

Chan LH, Edmond JM, Thompson G, Gillis K (1992) Lithium isotopic compositions of submarine basalts: implications for the lithium cycle in the oceans. Earth Planet Sci Lett 108:151–160

Chanton JP, Martens CS, Goldhaber MB (1987a) Biogeochemical cycling in an organic-rich coastal marine basin. 7. Sulfur mass balance, oxygen uptake and sulfide retention. Geochim Cosmochim Acta 51:1187–1199

Chanton JP, Martens CS, Goldhaber MB (1987b) Biogeochemical cycling in an organic-rich coastal marine basin. 8. A sulfur isotope budget balanced by differential diffusion across the sediment-water interface. Geochim Cosmochim Acta 51:1201–1208

Chaussidon M, Albarede F (1992) Secular boron isotope variations in the continental crust: an ion microprobe study. Earth Planet Sci Lett 108:229–241

Chaussidon M, Jambon A (1994) Boron content and isotopic composition of oceanic basalts: geochemical and cosmochemical implications. Earth Planet Sci Lett 121:277–291

Chaussidon M, Marty B (1995) Primitive boron isotope composition of the mantle. Science 269:383–386

Chaussidon M, Albarede F, Sheppard SMF (1987) Sulphur isotope heterogeneity in the mantle from ion microprobe measurements of sulphide inclusions in diamonds. Nature 330:242–244

Chaussidon M, Albarede F, Sheppard SMF (1989) Sulphur isotope variations in the mantle from ion microprobe analysis of microsulphide inclusions. Earth Planet Sci Lett 92:144–156

Chiba H, Chacko T, Clayton RN, Goldsmith JR (1989) Oxygen isotope fractionations involving diopside, forsterite, magnetite and calcite: application to geothermometry. Geochim Cosmochim Acta 53:2985–2995

Chivas AR, Andrew AS, Sinha AK, O'Neil JR (1982) Geochemistry of Pliocene-Pleistocene oceanic arc plutonic complex, Guadalcanal. Nature 300:139–143

Chung HM, Sackett WM (1979) Use of stable carbon isotope compositions of pyrolytically derived methane as maturity indices for carbonaceous materials. Geochim Cosmochim Acta 43:1979–1988

Ciais P, Tans PP, Trolier M, White JWC, Francey RJ (1995) A large northern hemisphere terrestrial CO_2 sink indicated by the $^{13}C/^{12}C$ ratio of atmospheric CO_2. Science 269:1098–1102

Cifuentes LA, Fogel ML, Pennock JR, Sharp JR (1989) Biogeochemical factors that influence the stable nitrogen isotope ratio of dissolved ammonium in the Delaware Estuary. Geochim Cosmochim Acta 53:2713–2721

Clausen P (1995) Lithium Isotopenmessungen – konventionell vs. Toatalevaporation und Integration. Dissertation der Math-Nat Fachbereiche, Universität Göttingen

Claypool GE, Holser WT, Kaplan IR, Sakai H, Zak I (1980) The age curves of sulfur and oxygen isotopes in marine sulfate and their mutual interpretation. Chem Geol 28:199–260

Clayton RN (1993) Oxygen isotopes in meteorites. Annu Rev Earth Planet Sci 21:115–149

Clayton RN, Epstein S (1958) The relationship between $^{18}O/^{16}O$ ratios in coexisting quartz, carbonate and iron oxides from various geological deposits. J Geol 66:352–373

Clayton RN, Mayeda TK (1963) The use of bromine pentafluoride in the extraction of oxygen from oxides and silicates for isotopic analysis. Geochim Cosmochim Acta 27:43–52

Clayton RN, Mayeda TK (1978) Genetic relations between iron and stony meteorites. Earth Planet Sci Lett 40:168–174

Clayton RN, Steiner A (1975) Oxygen isotope studies of the geothermal system at Warakei, New Zealand. Geochim Cosmochim Acta 39:1179–1186

Clayton RN, Friedman I, Graf DL, Mayeda TK, Meents WF, Shimp NF (1966) The origin of saline formation waters. 1. Isotopic composition. J Geophys Res 71:3869–3882

Clayton RN, Muffler LJP, White (1968) Oxygen isotope study of calcite and silicates of the River Branch No. I well, Salton Sea Geothermal Field, California. Am J Sci 266:968–979

Clayton RN, Grossman L, Mayeda TK (1973a) A component of primitive nuclear composition in carbonaceous meteorites. Science 182:485–488

Clayton RN, Hurd JM, Mayeda TK (1973b) Oxygen isotopic compositions of Apollo 15, 16 and 17 samples and their bearing on lunar origin and petrogenesis. Proc 4th Lunar Sci Conf Geochim Cosmochim Acta Suppl 2:1535–1542

Clayton RN, Goldsmith JR, Karel KJ, Mayeda TK, Newton RP (1975) Limits on the effect of pressure in isotopic fractionation. Geochim Cosmochim Acta 39:1197–1201

Clayton RN, Onuma N, Grossman C, Mayeda TK (1977) Distribution of the presolar component in Allende and other carbonaceous chondrites. Earth Planet Sci Lett 34:209–224

Clayton RN, Goldsmith JR, Mayeda TK (1989) Oxygen isotope fractionation in quartz, albite, anorthite and calcite. Geochim Cosmochim Acta 53:725–733

Cline JD, Kaplan IR (1975) Isotopic fractionation of dissolved nitrate during denitrification in the eastern tropical North Pacific Ocean. Mar Chem 3:271–299

Cloud PE, Friedman I, Sisler FD (1958) Microbiological fractionation of the hydrogen isotopes. Science 127:1394–1395

Cole DR, Ohmoto H, Lasaga AC (1983) Isotopic exchange in mineral fluid systems. I. Theoretical evaluation of oxygen isotopic reactions and diffusion. Geochim Cosmochim Acta 47:1681–1695

Coleman ML, Moore MP (1978) Direct reduction of sulfates to sulfur dioxide for isotope analysis. Anal Chem 50:1594–1595

Coleman ML, Sheppard TJ, Durham JJ, Rouse JE, Moore GR (1982) Reduction of water with zinc for hydrogen isotope analysis. Anal Chem 54:993–995

Connolly CA, Walter LM, Baadsgaard H, Longstaffe F (1990) Origin and evolution of formation fluids, Alberta Basin, western Canada sedimentary basin: II. Isotope systematics and fluid mixing. Appl Geochem 5:397–414

Cook N, Hoefs J (in press) Sulphur isotope characteristics of metamorphosed Cu-(Zn) volcanogenic massive sulphide deposits in the Norwegian Caledonides. Chem Geol

Coplen TB, Hanshaw BB (1973) Ultrafiltration by a compacted clay membrane. I. Oxygen and hydrogen isotopic fractionation. Geochim Cosmochim Acta 37:2295–2310

Coplen TB, Kendall C, Hopple J (1983) Comparison of stable isotope reference samples. Nature 302:236–238

Cortecci G, Longinelli A (1970) Isotopic composition of sulfate in rain water, Pisa, Italy. Earth Planet Sci Lett 8:36–40

Craig H (1953) The geochemistry of the stable carbon isotopes. Geochim Cosmochim Acta 3:53–92

Craig H (1957) Isotopic standards for carbon and oxygen and correction factors for mass-spectrometric analysis of carbon dioxide. Geochim Cosmochim Acta 12:133–149

Craig H (1961a) Isotopic variations in meteoric waters. Science 133:1702–1703

Craig H (1961b) Standard for reporting concentrations of deuterium and oxygen-18 in natural waters. Science 133:1833–1834

Craig H (1963) The isotopic geochemistry of water and carbon in geothermal areas. In: Nuclear geology of geothermal areas. Spoleto, 9–13 Sept 1963, pp 17–53

Craig H (1966) Isotopic composition and origin of the Red Sea and Salton Sea geothermal brines. Science 154:1544–1548

Craig H, Gordon L (1965) Deuterium and oxygen-18 variations in the ocean and the marine atmosphere. In: Symposium on marine geochemistry. Graduate School of Oceanography, Univ Rhode Island, Occ Publ No 3:277

Craig H, Keeling CD (1963) The effects of atmospheric N_2O on the measured isotopic composition of atmospheric CO_2. Geochim Cosmochim Acta 27:549–551

Craig H, Boato G, White DE (1956) Isotopic geochemistry of thermal waters. Proc 2nd Conf Nucl Process Geol Settings, p 29

Craig H, Chou CC, Welhan JA, Stevens CM, Engelkemeier A (1988) The isotopic composition of methane in polar ice cores. Science 242:1535–1539

Criss RE, Taylor HP (1986) Meteoric-hydrothermal systems. In: Stable isotopes in high temperature geological processes. Rev Miner 16:373–424

Criss RE, Champion DE, McIntyre DH (1985) Oxygen isotope, aeromagnetic and gravity anomalies associated with hydrothermally altered zones in the Yankee Fork Mining District, Custer County, Idaho. Econ Geol 80:1277–1296

Criss RE, Gregory RT, Taylor HP (1987) Kinetic theory of oxygen isotopic exchange between minerals and water. Geochim Cosmochim Acta 51:1099–1108

Criss RE, Fleck RJ, Taylor HP (1991) Tertiary meteoric hydrothermal systems and their relation to ore deposition, Northwestern United States and Southern British Columbia. J Geophys Res 96:13335–13356

Crowe DE, Valley JW, Baker KL (1990) Micro-analysis of sulfur isotope ratios and zonation by laser microprobe. Geochim Cosmochim Acta 54:2075–2092

Crowson RA, Showers WJ, Wright EK, Hoering TC (1991) Preparation of phosphate samples for oxygen isotope analysis. Anal Chem 63:2397–2400

Curry WB, Duplessy JC, Labeyrie LD, Shackleton NJ (1988) Quaternary deep-water circulation changes in the distribution of $\delta^{13}C$ of deep water ΣCO_2 between the last glaciation and the Holocene. Paleoceanography 3:317–342

Czamanske GK, Rye RO (1974) Experimentally determined sulfur isotope fractionations between sphalerite and galena in the temperature range 600 °C to 275 °C. Econ Geol 69:17–25

Dansgaard W (1964) Stable isotope in precipitation. Tellus 16:436–468

Dansgaard W, Johnsen SJ, Clausen HB (1993) Evidence for general instability of past climate from a 250 kyr ice-core record. Nature 364:218–220

Dean WE, Arthur MA, Claypool GE (1986) Depletion of ^{13}C in Cretaceous marine organic matter: source, diagenetic or environmental signal? Mar Geol 70:119–158

Degens ET, Epstein S (1962) Relationship between $^{18}O/^{16}O$ ratios in coexisting carbonates, cherts and diatomites. Bull Am Assoc Pet Geol 46:534–542

Degens ET, Guillard RRL, Sackett WM, Hellebust JA (1968a) Metabolic fractionation of carbon isotopes in marine plankton. I. Temperature and respiration experiments. Deep Sea Res 15:1–9

Degens ET, Behrendt M, Gotthardt B, Reppmann E (1968b) Metabolic fractionation of carbon isotopes in marine plankton. II. Data on samples collected off the coasts of Peru and Ecuador. Deep Sea Res 15:11–20

Deines P (1977) On the oxygen isotope distribution among mineral triplets in igneous and metamorphic rocks. Geochim Cosmochim Acta 41:1709–1730

Deines P (1980a) The carbon isotopic composition of diamonds: relationship to diamond shape, color, occurrence and vapor composition. Geochim Cosmochim Acta 44:943–962

Deines P (1980b) The isotopic composition of reduced organic carbon. In: Fritz P, Fontes JC (eds) Handbook of environmental geochemistry, vol 1. Elsevier, New York, pp 239–406

Deines P (1989) Stable isotope variations in carbonatites. In: Bell K (ed) Carbonatites, genesis and evolution. Hyman, London, 619pp

Deines P, Gold DP (1973) The isotopic composition of carbonatite and kimberlite carbonates and their bearing on the isotopic composition of deep-seated carbon. Geochim Cosmochim Acta 37:1709–1733

Deines P, Gurney JJ, Harris JW (1984) Associated chemical and carbon isotopic composition variations in diamonds from Finsch and Premier Kimberlite, South Africa. Geochim Cosmochim Acta 48:325–342

Deloule E, Albarede F, Sheppard SMF (1991) Hydrogen isotope heterogeneities in the mantle from ionprobe analysis of amphiboles from ultramafic rocks. Earth Planet Sci Lett 105:543–553

DeNiro MJ Epstein S (1977) Mechanism of carbon isotope fractionation associated with lipid synthesis. Science 197:261–263

DeNiro MJ, Epstein S (1978) Influence of diet on the distribution of carbon isotopes in animals. Geochim Cosmochim Acta 42:495–506

DeNiro MJ, Epstein S (1979) Relationship between the oxygen isotope ratios of terrestrial plant cellulose, carbon dioxide and water. Science 204:51–53

DeNiro MJ, Epstein S (1981) Isotopic composition of cellulose from aquatic organisms. Geochim Cosmochim Acta 45:1885–1894

Derry LA, Kaufmann AJ, Jacobsen SB (1992) Sedimentary cycling and environmental change in the Late Proterozoic: evidence from stable and radiogenic isotopes. Geochim Cosmochim Acta 56:1317–1329

Desaulniers DE, Kaufmann RS, Cherry JO, Bentley HW (1986) ^{37}Cl-^{35}Cl variations in a diffusion-controlled groundwater system. Geochim Cosmochim Acta 50:1757–1764

Des Marais DJ (1983) Light element geochemistry and spallogenesis in lunar rocks. Geochim Cosmochim Acta 47:1769–1781

Des Marais DJ, Moore JG (1984) Carbon and its isotopes in mid-oceanic basaltic glasses. Earth Planet Sci Lett 69:43–57

Deuser WG (1970) Extreme $^{13}C/^{12}C$ variations in Quaternary dolomites from the continental shelf. Earth Planet. Sci Lett 8:118–124

Deuser WG, Hunt JM (1969) Stable isotope ratios of dissolved inorganic carbon in the Atlantic. Deep Sea Res 16:221–225

Deutsch S, Ambach W, Eisner H (1966) Oxygen isotope study of snow and firn of an Alpine glacier. Earth Planet Sci Lett 1:197–201

Dickson JAD, Coleman ML (1980) Changes in carbon and oxygen isotope composition during limestone diagenesis. Sedimentology 27:107–118

Dickson JAD, Smalley PC, Raheim A, Stijfhoorn DE (1990) Intracrystalline carbon and oxygen isotope variations in calcite revealed by laser micro-sampling. Geology 18:809–811

Dipple GM, Ferry JM (1992) Fluid flow and stable isotope alteration in rocks at elevated temperatures with applications to metamorphism. Geochim Cosmochim Acta 56:3539–3550

Dobson PF, O'Neil JR (1987) Stable isotope composition and water contents of boninite series volcanic rocks from Chichi-jima, Bonin Islands, Japan. Earth Planet Sci Lett 82:75–86

Dobson PF, Epstein S, Stolper EM (1989) Hydrogen isotope fractionation between coexisting vapor and silicate glasses and melts at low pressure. Geochim Cosmochim Acta 53:2723–2730

Dodson MH (1973) Closure temperature in cooling geochronological and petrological systems. Contr Mineral Petrol 40:259–274

Dole M, Lange GA, Rudd DP, Zaukelies DA (1954) Isotopic composition of atmospheric oxygen and nitrogen. Geochim Cosmochim Acta 6:65–78

Donahue TM, Hoffman JH, Hodges RD, Watson AJ (1982) Venus was wet: a measurement of the ratio of deuterium to hydrogen. Science 216:630–633

Douthitt CB (1982) The geochemistry of the stable isotopes of silicon. Geochim Cosmochim Acta 46:1449–145

Dugan JP, Borthwick J, Harmon RS, Gagnier MA, Glahn JE, Kinsel EP, McLeod S, Viglino JA (1985) Guadinine hydrochloride method for determination of water oxygen isotope ratios and the oxygen-18 fractionation between carbon dioxide and water at 250C. Anal Chem 57:1734–1736

Duplessy JC, Shackleton NJ, Fairbanks RG, Labeyrie L, Oppo D, Kallel N (1988) Deepwater source variations during the last climatic cycle and their impact on the global circulation. Paleoceanography 3:343–360

Durka W, Schulze ED, Gebauer G, Voerkelius S (1994) Effects of forest decline on uptake and leaching of deposited nitrate determined from ^{15}N and ^{18}O measurements. Nature 372:765–767

Eadie BJ, Jeffrey LM (1973) $\delta^{13}C$ analyses of oceanic particulate organic matter. Mar Chem 1:199–209

Eastoe CJ, Guilbert JM (1992) Stable chlorine isotopes in hydrothermal processes. Geochim Cosmochim Acta 56:4247–4255

Eastoe CJ, Guilbert JM, Kaufmann RS (1989) Preliminary evidence for fractionation of stable chlorine isotopes in ore-forming hydrothermal deposits. Geology 17:285–288

Eggenkamp HGM (1994) $\delta^{37}Cl$: the geochemistry of chlorine isotopes. Thesis, University of Utrecht

Eiler JM, Baumgartner LP, Valley JW (1992) Intercrystalline stable isotope diffusion: a fast grain boundary model. Contr Mineral Petrol 112:543–557

Eiler JM, Valley JW, Baumgartner LP (1993) A new look at stable isotope thermometry. Geochim Cosmochim Acta 57:2571–2583

Eiler JM, Valley JW, Graham CM, Baumgartner LP (1995) Ion microprobe evidence for the mechanisms of stable isotope retrogression in high-grade metamorphic rocks. Contr Mineral Petrol 118:365–378

Eldridge CS, Compston W, Williams IS, Both RA, Walshe JL, Ohmoto H (1988) Sulfur isotope variability in sediment hosted massive sulfide deposits as determined using the ion microprobe SHRIMP. I. An example from the Rammelsberg ore body. Econ Geol 83:443–449

Eldridge CS, Compston W, Williams IS, Harris JW, Bristow JW (1991) Isotopic evidence for the involvement of recycled sediments in diamond formation. Nature 353:649–653

Eldridge CS, Williams IS, Walshe JL (1993) Sulfur isotope variability in sediment hosted massive sulfide deposits as determined using the ion microprobe SHRIMP. II. A study of the H.Y.C. deposit at McArthur River, Northern Territory, Australia. Econ Geol 88:1–26

Emiliani C (1955) Pleistocene temperatures. J Geol 63:538–578

Emrich K, Ehhalt DH, Vogel JC (1970) Carbon isotope fractionation during the precipitation of calcium carbonate. Earth Planet Sci Lett 8:363–371

Engel MH, Macko SA, Silfer JA (1990) Carbon isotope composition of individual amino acids in the Murchison meteorite. Nature 348:47–49

Epstein S, Taylor HP (1970) $^{18}O/^{16}O$, $^{30}Si/^{28}Si$, D/H and $^{13}C/^{12}C$ studies of lunar rocks and minerals. Science 167:533–535

Epstein S, Taylor HP (1971) $^{18}O/^{16}O$, $^{30}Si/^{28}Si$, D/H and $^{13}C/^{12}C$ ratios in lunar samples. Proc 2nd Lunar Sci Conf 2:1421–1441

Epstein S, Taylor HP (1972) $^{18}O/^{16}O$, $^{30}Si/^{28}Si$, $^{13}C/^{12}C$ and D/H studies of Apollo 14 and 15 samples. Proc 3rd Lunar Sci Conf 2:1429–1454

Epstein S, Yapp CJ, Hall JH (1976) The determination of the D/H ratio of non-exchangeable hydrogen in cellulose extracted from aquatic and land plants. Earth Planet Sci Lett 30:241–251

Epstein S, Thompson P, Yapp CJ (1977) Oxygen and hydrogen isotopic ratios in plant cellulose. Science 198:1209–1215

Epstein S, Krishnamurthy RV, Cronin JR, Pizzarello S, Yuen GU (1987) Unusual stable isotope ratios in amino acid and carboxylic acid extracts from the Murchison meteorite. Nature 326:477–479

Erez J, Luz B (1983) Experimental paleotemperature equation for planktonic foraminifera. Geochim Cosmochim Acta 47:1025–1031

Eslinger EV, Savin SM (1973) Oxygen isotope geothermometry of the burial metamorphic rocks of the Precambrian Belt Supergroup, Glacier National Park, Montana. Bull Geol Soc Am 84:2549–2560

Estep MF, Hoering TC (1980) Biogeochemistry of the stable hydrogen isotopes. Geochim Cosmochim Acta 44:1197–1206

Exley RA, Mattey DP, Boyd SR, Pillinger CT (1987) Nitrogen isotope geochemistry of basaltic glasses: implications for mantle degassing and structure. Earth Planet Sci Lett 81:163–174

Fairbanks RG (1989) A 17000 year glacio-eustatic sea level record: influence of glacial melting rates on the Younger Dryas event and deep ocean circulation. Nature 342:637–642

Farquhar GD, Lloyd J, Taylor JA (1993) Vegetation effects on the isotope composition of oxygen in atmospheric CO_2. Nature 363:439–443

Farquhar J, Chacko T, Frost BR (1993) Strategies for high temperature oxygen isotope thermometry: a worked example from the Laramie Anorthosite Complex, Wyoming, USA. Earth Planet Sci Lett 117:407–422

Farquhar J, Chacko T, Ellis DJ (1996) Preservation of oxygen isotopic compositions in granulites from Northwestern Canada and Enderby Land, Antarctica: implications for high-temperature isotopic thermometry. Contr Mineral Petrol (under review)

Ferrara G, Laurenzi MA, Taylor HP, Tonarini S, Turi B (1985) Oxygen and strontium isotope studies of K-rich volcanic rocks from the Alban Hills, Italy. Earth Planet Sci Lett 75:13–28

Ferrara G, Preite-Martinez M, Taylor HP, Tonarini S, Turi B (1986) Evidence for crustal assimilation, mixing of magmas, and a [87]Sr-rich upper mantle. An oxygen and strontium isotope study of the M. Vulsini volcanic area, Central Italy. Contrib Miner Petrol 92:269–280

Ferry JM (1992) Regional metamorphism of the Waits River Formation: delineation of a new type of giant hydrothermal system. J Petrol 33:45–94

Ferry JM, Dipple GM (1992) Models for coupled fluid flow, mineral reaction and isotopic alteration during contact metamorphism: the Notch Peak aureole, Utah. Am Mineral 77:577–591

Field CW, Gustafson LB (1976) Sulfur isotopes in the porphyry copper deposit at El Salvador, Chile. Econ Geol 71:1533–1548

Fiorentini E, Hoernes S, Hoffbauer R, Vitanage PW (1990) Nature and scale of fluid-rock exchange in granulite-grade rocks of Sri Lanka: a stable isotope study. In: Vielzeuf D, Vidal P (eds) Granulites and crustal evolution. Kluwer, Dordrecht, pp 311–338

Fogel ML, Cifuentes LA (1993) Isotope fractionation during primary production. In: Engel MH, Macko SA (eds) Organic geochemistry. Plenum Press, New York, pp 73–98

Fowler M, Harmon RS (1990) The oxygen isotope composition of lower crustal granulite xenoliths. In: Vielzeuf D (ed) Petrology and geochemistry of granulites. NATO/ASI Ser C 311:493–506

Francey RJ, Tans PP (1987) Latitudinal variation in oxygen-18 of atmospheric CO_2. Nature 327:495–497

Franchi IA, Wright IP, Pillinger CT (1993) Constraints on the formation conditions of iron meteorites based on concentrations and isotopic compositions of nitrogen. Geochim Cosmochim Acta 57:3105–3121

Frape SK, Fritz P (1987) Geochemical trends from groundwaters from the Canadian Shield. In: Fritz P, Frape SK (eds) Saline water and gases in crystalline rocks. Geol Assoc Can Spec Pap 33:19–38

Frape SK, Fritz P, McNutt RH (1984) Water-rock interaction and chemistry of groundwaters from the Canadian Shield. Geochim Cosmochim Acta 48:1617–1627

Freeman KH, Hayes JM, Trendel JM, Albrecht P (1990) Evidence from carbon isotope measurements for diverse origins of sedimentary hydrocarbons. Nature 343:254–256

Freyer HD (1979) On the ^{13}C-record in tree rings. I. [13]C variations in northern hemisphere trees during the last 150 years. Tellus 31:124–137

Freyer HD, Belacy N (1983) ^{13}C/^{12}C records in northern hemispheric trees during the past 500 years – anthropogenic impact and climatic superpositions. J Geophys Res 88:6844–6852

Fricke HC, Wickham SM, O'Neil JR (1992) Oxygen and hydrogen isotope evidence for meteoric water infiltration during mylonitization and uplift in the Ruby Mountains – East Humboldt Range core complex, Nevada. Contr Mineral Petrol 111:203–221

Friedman I (1953) Deuterium content of natural waters and other substances. Geochim Cosmochim Acta 4:89–103

Friedman I, O'Neil JR (1977) Compilation of stable isotope fractionation factors of geochemical interest. In: Fleischer M (ed) Data of geochemistry, Chap. KK, 6th edn. US Geol Survey Prof Pap 440KK

Friedman I, Scholz TG (1974) Isotopic composition of atmospheric hydrogen (1967–1969). J Geophys Res 79:785–788

Friedman I, O'Neil JR, Adami LH, Gleason JD, Hardcastle KG (1970) Water, hydrogen, deuterium, carbon-13, and oxygen-18 content of selected lunar material. Science 167:538–540

Friedman I, Hardcastle KG, Gleason JD (1974) Water and carbon in rusty lunar rock 66095. Science 185:346–349

Fry B (1988) Food web structure on Georges Bank from stable C, N and S isotopic compositions. Limnol Oceanogr 3:1182–1190

Fuex AN (1977) The use of stable carbon isotopes in hydrocarbon exploration. J Geochem Explor 7:155–188

Galimov EM (1985a) The biological fractionation of isotopes. Academic Press, New York

Galimov EM (1985b) The relation between formation conditions and variations in isotope compositions of diamonds. Geochem Int 22 (1):118–141

Galimov EM (1988) Sources and mechanisms of formation of gaseous hydrocarbons in sedimentary rocks. Chem Geol 71:77–95

Galimov EM (1991) Isotopic fractionation related to kimberlite magmatism and diamond formation. Geochim Cosmochim Acta 55:1697–1708

Ganor J, Matthews A, Paldor N (1989) Constraints on effective diffusivity during oxygen isotope exchange at a marble-schist contact, Sifnos (Cyclades), Greece. Earth Planet Sci Lett 94:208–216

Gao X, Thiemens MH (1991) Systematic study of sulfur isotopic composition in iron meteorites and the occurrence of excess ^{33}S and ^{36}S. Geochim Cosmochim Acta 55:2671–2681

Gao X, Thiemens MH (1993a) Isotopic composition and concentration of sulfur in carbonaceous chondrites. Geochim Cosmochim Acta 57:3159–3169

Gao X, Thiemens MH (1993b) Variations of the isotopic composition of sulfur in enstatite and ordinary chondrites. Geochim Cosmochim Acta 57:3171–3176

Garlick GD (1966) Oxygen isotope fractionation in igneous rocks. Earth Planet Sci Lett 1:361–368

Garlick GD (1969) The stable isotopes of oxygen. In: Wedepohl KH (ed) Handbook of geochemistry, vol 8B. Springer, Berlin Heidelberg New York

Garlick GD, Epstein S (1967) Oxygen isotope ratios in coexisting minerals of regionally metamorphosed rocks. Geochim Cosmochim Acta 31:181

Gat JR (1971) Comments on the stable isotope method in regional groundwater investigation. Water Resource Res 7:980

Gat JR (1984) The stable isotope composition of Dead Sea waters. Earth Planet Sci Lett 71:361–376

Gat JR, Issar A (1974) Desert isotope hydrology: water sources of the Sinai desert. Geochim Cosmochim Acta 38:1117–11131

Gautier DL (1982) Siderite concretions: indicators of early diagenesis in the Ganimon Shale (Cretaceous). J Sediment Petrol 52:859–871

Gerdes ML, Baumgartner LP, Person M, Rumble D (1995) One- and two-dimensional models of fluid flow and stable isotope exchange at an outcrop in the Adamello contact aureole, Southern Alps, Italy. Am Miner 80:1004–1019

Gerlach TM, Taylor BE (1990) Carbon isotope constraints on degassing of carbon dioxide from Kilauea volcano. Geochim Cosmochim Acta 54:2051–2058

Gerlach TM, Thomas DM (1986) Carbon and sulphur isotopic composition of Kilauea parental magma. Nature 319:480–483

Giggenbach WF (1982) Carbon-13 exchange between CO_2 and CH_4 under geothermal conditions. Geochim Cosmochim Acta 46:159–165

Giggenbach WF (1992) Isotopic shifts in waters from geothermal and volcanic systems along convergent plate boundaries and their origin. Earth Planet Sci Lett 113:495–510

Giletti BJ (1985) The nature of oxygen transport within minerals in the presence of hydrothermal water and the role of diffusion. Chem Geol 53:197–206

Giletti BJ (1986) Diffusion effect on oxygen isotope temperatures of slowly cooled igneous and metamorphic rocks. Earth Planet Sci Lett 77:218–228

Girard JP, Aronson JL, Savin SM (1988) Separation, K-Ar dating and $^{18}O/^{16}O$ measurements of diagenetic K-feldspar overgrowths: an example from the Lower Cretaceous arkoses of the Angola Margin. Geochim Cosmochim Acta 52:2207–2214

Given RK, Lohmann KC (1985) Derivation of the original isotopic composition of Permian marine cements. J Sediment Petrol 55:430–439

Godfrey JD (1962) The deuterium content of hydrous minerals from the East Central Sierra Nevada and Yosemite National Park. Geochim Cosmochim Acta 26:1215–1245

Goldhaber MB, Kaplan IR (1974) The sedimentary sulfur cycle. In: Goldberg EB (ed) The sea, vol. IV. Wiley, New York

Gonfiantini R (1978) Standards for stable isotope measurements in natural compounds. Nature 271:534–536

Gonfiantini R (1984) Advisory group meeting on stable isotope reference samples for geochemical and hydrological investigations. Rep Director General IAEA Vienna

Gonfiantini R (1986) Environmental isotopes in lake studies. In: Fritz P, Fontes J (eds) Handbook of environmental isotope geochemistry, vol 2. Elsevier, Amsterdam, pp 112–168

Gonfiantini R, Fontes JC (1963) Oxygen isotope fractionation in the water of crystallization of gypsum. Nature 200:644–646

Grady MM, Wright IP, Swart PK, Pillinger CT (1985) The carbon and nitrogen isotopic composition of ureilites: implications for their genesis. Geochim Cosmochim Acta 49:903–915

Graham AM, Graham CM, Harmon RS (1982) Origins of mantle waters: stable isotope evidence from amphibole-bearing plutonic cumulate blocks in calc-alkaline volcanics, Grenada, Lesser Antilles. Abstr 5th Int Conf Geochronology, Cosmochronology, Isotope Geology, pp 119–120

Graham CM (1981) Experimental hydrogen isotope studies. III. Diffusion of hydrogen in hydrous minerals and stable isotope exchange in metamorphic rocks. Contrib Miner Petrol 76:216–228

Graham CM, Sheppard SMF, Heaton THE (1980) Experimental hydrogen isotope studies. I. Systematics of hydrogen isotope fractionation in the systems epidote-H_2O, zoisite-H_2O and $AlO(OH)$-H_2O. Geochim Cosmochim Acta 44:353–364

Graham CM, Harmon RS, Sheppard SMF (1984) Experimental hydrogen isotope studies: hydrogen isotope exchange between amphibole and water. Am Miner 69:128–138

Green GR, Ohmoto D, Date J, Takahashi T (1983) Whole-rock oxygen isotope distribution in the Fukazawa-Kosaka Area, Hokuroko District, Japan and its potential application to mineral exploration . Econ Geol Monogr 5:395–411

Gregory RT, Criss RE (1986) Isotopic exchange in open and closed systems. Rev Mineral 16:91–127

Gregory RT, Taylor HP (1981) An oxygen isotope profile in a section of Cretaceous oceanic crust, Samail Ophiolite, Oman: evidence for $\delta^{18}O$ buffering of the oceans by deep (>5 km) seawater-hydrothermal circulation at Mid-Ocean Ridges. J Geophys Res 86:2737–2755

Gregory RT, Taylor HP (1986a) Possible non-equilibrium oxygen isotope effects in mantle nodules, an alternative to the Kyser-O'Neil-Carmichael $^{18}O/^{16}O$ geothermometer. Contr Mineral Petrol 93:114–119

Gregory RT, Taylor HP (1986b) Non-equilibrium, metasomatic $^{18}O/^{16}O$ effects in upper mantle mineral assemblages. Contr Mineral Petrol 93:124–135

Gregory RT, Criss RE, Taylor HP (1989) Oxygen isotope exchange kinetics of mineral pairs in closed and open systems: applications to problems of hydrothermal alteration of igneous rocks and Precambrian Iron Formations. Chem Geol 75:1–42

Grootes PM, Stuiver M, White JWC, Johnsen S, Jouzel J (1993) Comparison of oxygen isotope records from the GISP-2 and GRIP Greenland ice cores. Nature 366:552–554

Grossman EL (1984) Carbon isotopic fractionation in live benthic foraminifera – comparison with inorganic precipitate studies. Geochim Cosmochim Acta 48:1505–1512

Grunder A, Wickham SM (1991) Homogenization and lowering of $^{18}O/^{16}O$ in mid-crustal rocks during extension related magmatism in Eastern Nevada. Earth Planet Sci Lett 107:416–431

Haack U, Hoefs J, Gohn E (1982) Constraints on the origin of Damaran granites by Rb/Sr and $\delta^{18}O$ data. Contrib Mineral Petrol 79:279–289

Hackley KC, Anderson TF (1986) Sulfur isotopic variations in low-sulfur coals from the Rocky Mountain region. Geochim Cosmochim Acta 50:703–1713

Haendel D, Mühle K, Nitzsche HIM, Stiehl G, Wand U (1986) Isotopic variations of the fixed nitrogen in metamorphic rocks. Geochim Cosmochim Acta 50:749–758

Hagemann R, Nief G, Roth E (1970) Absolute isotopic scale for deuterium analysis of natural waters. Absolute D/H ratio for SMOW. Tellus 22:712–715

Haimson M, Knauth LP (1983) Stepwise fluorination – a useful approach for the isotopic analysis of hydrous minerals. Geochim Cosmochim Acta 47:1589–1595

Halbout J, Robert F, Javoy M (1990) Hydrogen and oxygen isotope compositions in kerogen from the Orgueil meteorite: clues to a solar origin. Geochim Cosmochim Acta 54:1453–1462

Haines EB (1976) Relation between the stable carbon isotope composition of fiddler crabs, plants and soils in a salt marsh. Limnol Oceanogr 21:880–883

Hardie LA (1987) Dolomitization – a critical view of some current views. J Sed Petrol 57:166–183

Harmon RS, Hoefs J (1984) O-isotope relationships in Cenozoic volcanic rocks: evidence for a heterogeneous mantle source and open-system magma genesis. In: Proc ISEM Field Conf Open Magmatic System, pp 69–71

Harmon RS, Hoefs J (1986) S-isotope relationships in Late Cenozoic destructive plate margin and continental intraplate volcanic rocks. Terra Cognita 6:182

Harmon RS, Hoefs J (1995) Oxygen isotope heterogeneity of the mantle deduced from global ^{18}O systematics of basalts from different geotectonic settings. Contr Mineral Petrol 120: 95–114

Harmon RS, Hoefs J, Wedepohl KH (1987) Stable isotope (O,H,S) relationships in Tertiary basalts and their mantle xenoliths from the Northern Hessian Depression, W Germany. Contr Mineral Petrol 95:350–369

Harrison AG, Thode HG (1957a) Kinetic isotope effect in chemical reduction of sulphate. Faraday Soc Trans 53:1648–1651

Harrison AG, Thode HG (1957b) Mechanism of the bacterial reduction of sulphate from isotope fractionation studies. Faraday Soc Trans 54:84–92

Harte B, Otter M (1992) Carbon isotope measurements on diamonds. Chem Geol 101:177–183

Hartmann M, Nielsen H (1969) $\delta^{34}S$-Werte in rezenten Meeressedimenten und ihre Deutung am Beispiel einiger Sedimentprofile aus der westlichen Ostsee. Geol Rundsch 58:621–655

Hayes JM (1983) Practice and principles of isotopic measurements in organic geochemistry. In: Organic geochemistry of contemporaneous and ancient sediments, Great Lakes Section, SEPM, Bloomington, Ind., pp 5.1–5.31

Hayes JM, Kaplan IR, Wedeking KW (1983) Precambrian organic chemistry, preservation of the record. In: Schopf JW (ed) Earth's earliest biosphere: its origin and evolution, chap 5. Princeton University Press, Princeton, pp 93–132

Hayes JM, Popp BN, Takigiku R, Johnson MW (1989) An isotopic study of biogeochemical relationships between carbonates and organic carbon in the Greenhorn Formation. Geochim Cosmochim Acta 53:2961–2972

Hays JD, Imbrie J, Shackleton NJ (1976) Variations in the earth's orbit: pacemaker of the ice ages. Science 194:943–954

Heaton THE (1986) Isotopic studies of nitrogen pollution in the hydrosphere and atmosphere: a review. Chem Geol 59:87–102

Hedenquist JW, Lowenstern JB (1994) The role of magmas in the formation of hydrothermal ore deposits. Nature 370:519–527

Heidenreich JE, Thiemens MH (1983) A non-mass-dependent isotope effect in the production of ozone from molecular oxygen. J Chem Phys 78:892–895

Heidenreich JE, Thiemens MH (1985) The non-mass-dependent oxygen isotope effect in the electro dissociation of carbon dioxide: a step toward understanding NoMaD chemistry. Geochim Cosmochim Acta 49:1303–1306

Hemming NG, Hanson GN (1992) Boron isotopic composition in modern marine carbonates. Geochim Cosmochim Acta 56:537–543

Hitchon B, Friedman I (1969) Geochemistry and origin of formation waters in the western Canada sedimentary basin. 1. Stable isotopes of hydrogen and oxygen. Geochim Cosmochim Acta 33:1321–1349

Hitchon B, Krouse HR (1972) Hydrogeochemistry of the surface waters of the Mackenzie River drainage basin, Canada. III. Stable isotopes of oxygen, carbon and sulfur. Geochim Cosmochim Acta 36:1337–1357

Hoefs J (1970) Kohlenstoff- und Sauerstoff-Isotopenuntersuchungen an Karbonatkonkretionen und umgebendem Gestein. Contrib Mineral Petrol 27:66–79

Hoefs J (1981) Isotopic composition of the ocean-atmospheric system in the geologic past. In: O'Connell RJ, Fyfe WS (eds) Evolution of the earth. Geodynamic Series, vol 5. American Geophysical Union, Washington, DC, pp 110–119

Hoefs J (1992) The stable isotope composition of sedimentary iron oxides with special reference to Banded Iron Formations. In: Clauer N, Chaudhuri S (eds) Isotopic signatures and sedimentary records. Springer, Berlin Heidelberg New York, pp 199–213 (Lecture notes in earth science, vol 43)

Hoefs J, Emmermann R (1983) The oxygen isotope composition of Hercynian granites and pre-Hercynian gneisses from the Schwarzwald, SW Germany. Contrib Mineral Petrol 83:320–329

Hoering TC (1955) Variations in nitrogen-15 abundance in naturally occurring substances. Science 122:1233

Hoering T (1956) Variations in the nitrogen isotope abundance. Proc 2nd Conf Nucl Process Geol Settings, p 85

Hoering T (1975) The biochemistry of the stable hydrogen isotopes. Carnegie Inst Washington Yearb 74:598

Hoering T, Parker PL (1961) The geochemistry of the stable isotopes of chlorine. Geochim Cosmochim Acta 23:186–199

Hoernes S, Van Reenen DC (1992) The oxygen isotopic composition of granulites and retrogressed granulites from the Limpopo Belt as a monitor of fluid-rock interaction. Precambr Res 55:353–364

Hoffman JH, Hodges RR, McElroy MB, Donahue TM, Kolpin M (1979) Composition and structure of the Venus atmosphere: results from Pioneer Venus. Science 205:49–52

Holser WT (1977) Catastrophic chemical events in the history of the ocean. Nature 267:403–408

Holser WT, Kaplan IR (1966) Isotope geochemistry of sedimentary sulfates. Chem Geol 1:93–135

Holser WT, Magaritz M (1992) Cretaceous/Tertiary and Permian/Triassic boundary events compared. Geochim Cosmochim Acta 56:3297–3309

Holser WT, Kaplan IR, Sakai H, Zak I (1979) Isotope geochemistry of oxygen in the sedimentary sulfate cycle. Chem Geol 25:1–17

Holser WT, Schidlowski M, MacKenzie FT, Maynard JB (1988) Geochemical cycles of carbon and sulfur. In: Gregor CB, Garrels RM, Mackenzie FT (eds) Geochemical cycles in the evolution of the earth. Wiley, New York, pp 105–173

Holt BD, Engelkemeier AG (1970) Thermal decomposition of barium sulfate to sulfur dioxide for mass spectrometric analysis. Anal Chem 42:1451–1453

Horibe Y, Shigehara K, Takakuwa Y (1973) Isotope separation factors of carbon dioxide-water system and isotopic composition of atmospheric oxygen. J Geophys Res 78:2625–2629

Horita J (1989) Stable isotope fractionation factors of water in hydrated salt minerals. Earth Planet Sci Lett 95:173–179

Horita J, Wesolowski DJ (1994) Liquid-vapor fractionation of oxygen and hydrogen isotopes of water from the freezing to the critical temperature. Geochim Cosmochim Acta 58:3425–3437

Horita J, Wesolowski DJ, Cole DR (1993) The activity-composition relationship of oxygen and hydrogen isotopes in aqueous salt solutions. I. Vapor-liquid water equilibration of single salt solutions from 50 to 100 °C. Geochim Cosmochim Acta 57:2797–2817

Hudson JD (1977) Stable isotopes and limestone lithification. J Geol Soc (Lond) 133:637–660

Hulston JR (1977) Isotope work applied to geothermal systems at the Institute of Nuclear Sciences, New Zealand. Geothermics 5:89–96

Hulston JR (1978) Methods of calculating isotopic fractionation in minerals. In: Stable isotopes in the earth sciences. DSIR Bull 220:211–219

Hulston JR, Thode HG (1965) Variations in the ^{33}S, ^{34}S and ^{36}S contents of meteorites and their relations to chemical and nuclear effects. J Geophys Res 70: 3475–3484

Ionov DA, Hoefs J, Wedepohl KH, Wiechert U (1992) Contents and isotopic composition of sulfur in ultramafic xenoliths from Central Asia. Earth Planet Sci Lett 111:269–286

Irwin H, Coleman M, Curtis C (1977) Isotopic evidence for the source of diagenetic carbonate during burial of organic-rich sediments. Nature 269:209–213

Ishibashi J, Sano Y, Wakita H, Gamo T, Tsutsumi M, Sakai H (1995) Helium and carbon geochemistry of hydrothermal fluids from the Mid-Okinawa trough back arc basin, southwest of Japan. Chem Geology 123:1–15

Ishikawa T, Nakamura E (1993) Boron isotope systematics of marine sediments. Earth Planet Sci Lett 117:567–580

Iyer SS, Barbosa JSF, Choudhuri A, Krouse HR (1995) Possible sources of CO_2 in granulites: carbon isotope evidence from the Jequie Complex, Brazil. Petrology 3:226–237

James AT (1983) Correlation of natural gas by use of carbon isotopic distribution between hydrocarbon components. Am Assoc Petrol Geol Bull 67:1167–1191

James AT (1990) Correlation of reservoired gases using the carbon isotopic compositions of wet gas components. Am Assoc Petrol Geol Bull 74:1441–1458

James DE (1981) The combined use of oxygen and radiogenic isotopes as indicators of crustal contamination. Annu Rev Earth Planet Sci 9:311–344

Jamtveit B, Hervig RL (1994) Constraints on transport and kinetics in hydrothermal systems from zoned garnet crystals. Science 263:505–508

Jasper JP, Hayes JM (1990) A carbon isotope record of CO_2 levels during the late Quaternary. Nature 347:462–464

Javoy M (1977) Stable isotopes and geothermometry. J Geol Soc 133:609–636

Javoy M, Pineau F (1991) The volatile record of a „popping" rock from the Mid-Atlantic Ridge at 140N: Chemical and isotopic composition of gas trapped in the vesicles. Earth Planet Sci Lett 107:598–611

Javoy M, Pineau F, Delorme H (1986) Carbon and nitrogen isotopes in the mantle. Chem Geol 57:41–62

Jeffrey AW, Pflaum RC, Brooks JM, Sackett WM (1983) Vertical trends in particulate organic carbon $^{13}C/^{12}C$ ratios in the upper water column. Deep Sea Res 30:971–983

Jenden PD, Kaplan IR, Poreda RJ, Craig H (1988) Origin of nitrogen-rich natural gases in the California Great Valley: evidence from helium, carbon and nitrogen isotope ratios. Geochim Cosmochim Acta 52:851–861

Jensen ML, Nakai N (1962) Sulfur isotope meteorite standards, results and recommendations. In: Jensen ML (ed) Biogeochemistry of sulfur isotopes. NSF Symp Vol, p 31

Jiang J, Clayton RN, Newton RC (1988) Fluids in granulite facies metamorphism: a comparative oxygen isotope study on the South India and Adirondack high grade terrains. J Geol 96:517–533

Jiang SY, Palmer MR, Ding TP, Wan DF (1994) Silicon isotope geochemistry of the Sullivan Pb-Zn deposit, Canada: a preliminary study. Econ Geol 89:1623–1629

Jones HD et al. (1996) Sulfur isotope geochemistry of southern Appalachian Mississippi Valley-type deposits. Econ Geol (in press)

Jouzel J, Merlivat L, Roth E (1975) Isotopic study of hail. J Geophys Res 80:5015–5030

Jouzel J, Lorius C, Petit JR, Barkov NI, Kotlyakov VM, Petrow VM (1987) Vostok ice core: a continuous isotopic temperature record over the last climatic cycle (160|000 years). Nature 329:403–408

Jouzel J, Barkov NI, Barbola JM (1993) Extending the Vostok ice-core record of palaeoclimate to the penultimate glacial period. Nature 364:407–412

Junk G, Svec H (1958) The absolute abundance of the nitrogen isotopes in the atmosphere and compressed gas from various sources. Geochim Cosmochim Acta 14:234–243

Kakihana H, Kotaka M, Shohei S, Nomura M, Okamoto N (1977) Fundamental studies on the ion-exchange separation of boron isotopes. Bull Chem Soc Jpn 50:158–163

Kaplan IR (1975) Stable isotopes as a guide to biogeochemical processes. Proc R Soc Lond [B] 189:183–211

Kaplan IR (1983) Stable isotopes of sulfur, nitrogen and deuterium in recent marine environments. In: Stable isotopes in sedimentary geology. SEPM short course 10, Dallas

Kaplan IR, Hulston JR (1966) The isotopic abundance and content of sulfur in meteorites. Geochim Cosmochim Acta 30:479–496

Kaplan IR, Rittenberg SC (1964) Microbiological fractionation of sulphur isotopes. J Gen Microbiol 34:195–212

Kaplan IR, Rafter TA, Hulston JR (1960) Sulphur isotope variations in nature: application to some biogeochemical problems. NZ J Sci 3:338

Kaplan IR, Emery KO, Rittenberg SC (1963) The distribution and isotopic abundance of sulphur in recent marine sediments off Southern California. Geochim Cosmochim Acta 27:297–332

Karhu J, Epstein S (1986) The implications of the oxygen isotope records in coexisting cherts and phosphates. Geochim Cosmochim Acta 50:1745–1756

Kaufmann RS, Long A, Bentley H, Davis S (1984) Natural chlorine isotope variations. Nature 309:338–340

Kaufmann RS, Long A, Bentley H, Campbell DJ (1986) Chlorine isotope distribution of formation water in Texas and Louisiana. Bull Am Assoc Petrol Geol 72:839–844

Kaye J (1987) Mechanisms and observations for isotope fractionation of molecular species in planetary atmospheres. Rev Geophys 25:1609–1658

Keeling CD (1958) The concentration and isotopic abundance of atmospheric carbon dioxide in rural areas. Geochim Cosmochim Acta 13:322–334

Keeling CD (1961) The concentration and isotopic abundances of carbon dioxide in rural and marine air. Geochim Cosmochim Acta 24:277–298

Keeling CD, Mook WG, Tans P (1979) Recent trends in the $^{13}C/^{12}C$ ratio of atmospheric carbon dioxide. Nature 277:121–123

Keeling CD, Carter AF, Mook WG (1984) Seasonal, latitudinal and secular variations in the abundance and isotopic ratio of atmospheric carbon dioxide. II. Results from oceanographic cruises in the tropical Pacific Ocean. J Geophys Res 89:4615–4628

Keeling CD, Bacastow RB, Carter AF, Piper SC, Whorf TR, Heimann M, Mook WG, Roeloffzen H (1989) A three dimensional model of atmospheric CO_2 transport based on observed winds. 1. Analysis of observational data. Geophys Monogr 55:165–236

Keeling CD, Whorf TP, Wahlen M, van der Plicht J (1995) Interannual extremes in the rate of rise of atmospheric carbon dioxide since 1980. Nature 375:666–670

Keith ML, Weber JN (1964) Carbon and oxygen isotopic composition of selected limestones and fossils. Geochim Cosmochim Acta 28:1787–1816

Kelley SP, Fallick AE (1990) High precision spatially resolved analysis of $\delta^{34}S$ in sulphides using a laser extraction technique. Geochim Cosmochim Acta 54:883–888

Kelly WC, Rye RO, Livnat A (1986) Saline minewaters of the Keweenaw Peninsula, Northern Michigan: their nature, origin and relation to similar deep waters in Precambrian crystalline rocks of the Canadian Shield. Am J Sci 286:281–308

Kelts K, McKenzie JA (1982) Diagenetic dolomite formation in Quaternary anoxic diatomaceous muds of DSDP Leg 64, Gulf of California. Initial Rep DSDP 64:553–569

Kemp ALW, Thode HG (1968) The mechanism of the bacterial reduction of sulphate and of sulphite from isotopic fractionation studies. Geochim Cosmochim Acta 32:71–91

Kempton PD, Harmon RS (1992) Oxygen isotope evidence for large-scale hybridization of the lower crust during magmatic underplating. Geochim Cosmochim Acta 56:971–986

Kendall C, Grim E (1990) Combustion tube method for measurement of nitrogen isotope ratios using calcium oxide for total removal of carbon dioxide and water. Anal Chem 62:526–529

Kenig F, Hayes JM, Popp BN, Summons RE (1994) Isotopic biogeochemistry of the Oxford Clay Formation (Jurassic), UK. J Geol Soc (Lond) 151:139–152

Kennicutt MC, Barker C, Brooks JM, De Freitaas DA, Zhu GH (1987) Selected organic matter indicators in the Orinoco, Nile and Changjiang deltas. Org Geochem 11:41–51

Kerridge JF (1983) Isotopic composition of carbonaceous-chondrite kerogen: evidence for an interstellar origin of organic matter in meteorites. Earth Planet Sci Lett 64:186–200

Kerridge JF (1993) Long-term compositional variation in solar corpuscular radiation: evidence from nitrogen isotopes in the lunar regolith. Rev Geophys 31:423–437

Kerridge JF, Haymon RM, Kastner M (1983) Sulfur isotope systematics at the 21oN site, East Pacific Rise. Earth Planet Sci Lett 66:91–100

Kerridge JF, Chang S, Shipp R (1987) Isotopic characterization of kerogen-like material in the Murchson carbonaceous chondrite. Geochim Cosmochim Acta 51:2527–2540

Kerrich R, Rehrig W (1987) Fluid motion associated with Tertiary mylonitization and detachment faulting: $^{18}O/^{16}O$ evidence from the Picacho metamorphic core complex, Arizona. Geology 15:58–62

Kerrich R, Latour TE, Willmore L (1984) Fluid participation in deep fault zones: evidence from geological, geochemical and to $^{18}O/^{16}O$ relations. J Geophys Res 89:4331–4343

Kharaka YK, Berry FAF, Friedman I (1974) Isotopic composition of oil-field brines from Kettleman North Dome, California and their geologic implications. Geochim Cosmochim Acta 37:1899–1908

Kieffer SW (1982) Thermodynamic and lattice vibrations of minerals: 5. Application to phase equilibria, isotopic fractionation and high-pressure thermodynamic properties. Rev Geophys Space Phys 20:827–849

Kim KR, Craig H (1990) Two isotope characterization of N_2O in the Pacific Ocean and constraints on its origin in deep water. Nature 347:58–61

Kim KR, Craig H (1993) Nitrogen-15 and oxygen-18 characteristics of nitrous oxide. Science 262:1855–1858

Kirkley MB, Gurney JJ, Otter ML, Hill SJ, Daniels LR (1991) The application of C-isotope measurements to the identification of the sources of C in diamonds. Appl Geochem 6:477–494

Kirschenbaum I, Smith JS, Crowell T, Graff J, McKee R (1947) Separation of the nitrogen isotopes by the exchange reaction between ammonia and solutions of ammonium nitrate. J Chem Phys 15:440–446

Kitchen NE, Valley JW (1995) Carbon isotope thermometry in marbles of the Adirondack Mountains, New York. J Metamorph Geol 13:577–594

Knauth LP (1988) Origin and mixing history of brines, Palo Duro Basin, Texas, USA. Appl Geochem 3:455–474

Knauth LP, Beeunas MA (1986) Isotope geochemistry of fluid inclusions in Permian halite with implications for the isotopic history of ocean water and the origin of saline formation waters. Geochim Cosmochim Acta 50:419–433

Knauth LP, Lowe DR (1978) Oxygen isotope geochemistry of cherts from the Onverwacht group (3.4 billion years), Transvaal, South Africa, with implications for secular variations in the isotopic composition of chert. Earth Planet Sci Lett 41:209–222

Knoll AH, Hayes JM, Kaufman AJ, Swett K, Lambert IB (1986) Secular variation in carbon isotope ratios from Upper Proterozoic successions of Svalbard and East Greenland. Nature 321:832–838

Kohn MJ (1993) Modeling of prograde mineral $\delta^{18}O$ changes in metamorphic systems. Contr Mineral Petrol 113:249–261

Kolodny Y, Kerridge JF, Kaplan IR (1980) Deuterium in carbonaceous chondrites. Earth Planet Sci Lett 46:149–153

Kolodny Y, Luz B, Navon O (1983) Oxygen isotope variations in phosphate of biogenic apatites, I. Fish bone apatite – rechecking the rules of the game. Earth Planet Sci Lett 64:393–404

Krichevsky MI, Sesler FD, Friedman I, Newell M (1961) Deuterium fractionation during molecular H_2 formation in a marine pseudomonad. J Biol Chem 236:2520

Krishnamurthy RV, Epstein S, Cronin JR, Pizzarello S, Yuen GU (1992) Isotopic and molecular analyses of hydrocarbons and monocarboxylic acids of the Murchison meteorite. Geochim Cosmochim Acta 56:4045–4058

Kroopnick P (1975) Respiration, photosynthesis, and oxygen isotope fractionation in oceanic surface water. Limnol Oceanogr 20:988–992

Kroopnick P (1985) The distribution of ^{13}C of ΣCO_2 in the world oceans. Deep Sea Res 32:57–84

Kroopnick P, Craig H (1972) Atmospheric oxygen: isotopic composition and solubility fractionation. Science 175:54–55

Kroopnick P, Craig H (1976) Oxygen isotope fractionation in dissolved oxygen in the deep sea. Earth Planet Sci Lett 32:375–388

Kroopnick P, Weiss RF, Craig H (1972) Total CO_2, ^{13}C and dissolved oxygen-^{18}O at Geosecs II in the North Atlantic. Earth Planet Sci Lett 16:103–110

Krouse HR, Case JW (1983) Sulphur isotope abundances in the environment and their relation to long term sour gas flaring, near Valleyview, Alberta. Final Report Res Management Division, University Alberta RMD Rep 83/18

Krouse HR, Grinenko VA (1991) Stable isotopes in the assessment of natural and anthropogenic sulphur in the environment. SCOPE 1991, J Wiley and Sons

Krouse HR, Viau CA, Eliuk LS, Ueda A, Halas S (1988) Chemical and isotopic evidence of thermochemical sulfate reduction by light hydrocarbon gases in deep carbonate reservoirs. Nature 333:415–419

Kung CC, Clayton RN (1978) Nitrogen abundances and isotopic compositions in stony meteorites. Earth Planet Sci Lett 38:421–435

Kyser TK (1995) Micro-analytical techniques in stable isotope geochemistry. Can Mineral 33:261–278

Kyser TK, O'Neil JR (1984) Hydrogen isotope systematics of submarine basalts. Geochim Cosmochim Acta 48:2123–2134

Kyser TK, O'Neil JR, Carmichael ISE (1981) Oxygen isotope thermometry of basic lavas and mantle nodules. Contrib Mineral Petrol 77:11–23

Kyser TK, O'Neil JR, Carmichael ISE (1982) Genetic relations among basic lavas and mantle nodules. Contrib Mineral Petrol 81:88–102

Kyser TK, O'Neil JR, Carmichael ISE (1986) Reply to „Possible non-equilibrium oxygen isotope effects in mantle nodules, an alternative to the Kyser-O'Neil-Carmichael geothermometer. Contr Mineral Petrol 93:120–123

Labeyrie LD, Juillet A (1982) Oxygen isotope exchangeability of diatom valve silica; interpretation and consequences for paleoclimatic studies. Geochim Cosmochim Acta 46:967–975

Labeyrie LD, Duplessy JC, Blanc PL (1987) Deep water formation and temperature variations over the last 125|000 years. Nature 327:477–482

Lancet MS, Anders E (1970) Carbon isotope fractionation in the Fischer-Tropsch synthesis and in meteorites. Science 170:980–982

Land LS (1980) The isotopic and trace element geochemistry of dolomite: the state of the art. In: Concepts and models of dolomitization. Soc Econ Paleontol Min Spec Publ 28:87–110

Land LS (1995) Comment on „Oxygen and carbon isotopic composition of Ordovician brachiopods: implications for coeval seawater" by H Quing and J Veizer. Geochim Cosmochim Acta 59:2843–2844

Lane GA, Dole M (1956) Fractionation of oxygen isotopes during respiration. Science 123:574–576

Lawrence JR (1989) The stable isotope geochemistry of deep-sea pore water. Elsevier, Amsterdam, pp 317–356 (Handbook of environmental isotope geochemistry, vol 3)

Lawrence JR, Gieskes JM (1981) Constraints on water transport and alteration in the oceanic crust from the isotopic composition of the pore water. J Geophys Res 86: 7924–7934

Lawrence JR, Raskes-Meaux J (1993) The stable isotopic composition of ancient kaolinites of North America. In: Climate change in continental isotopic records. Geophys Monogr 78:249–261

Lawrence JR, Taviani M (1988) Extreme hydrogen, oxygen and carbon isotope anomalies in the pore waters and carbonates of the sediments and basalts from the Norwegian Sea: methane and hydrogen from the mantle? Geochim Cosmochim Acta 52:2077–2083

Lawrence JR, Taylor HP (1971) Deuterium and oxygen-18 correlation: clay minerals and hydroxides in Quaternary soils compared to meteoric waters. Geochim Cosmochim Acta 35:993–1003

Lawrence JR, Taylor HP (1972) Hydrogen and oxygen isotope systematics in weathering profiles. Geochim Cosmochim Acta 36:1377–1393

Lawrence JR, White JWC (1991) The elusive climate signal in the isotopic composition of precipitation. In: Stable isotope geochemistry: a tribute to Samuel Epstein. Geochemical Soc Spec Publi 3:169–185

Lawrence JR, Gieskes JM, Broecker WS (1975) Oxygen isotope and cation composition of DSDP pore waters and the alteration of layer II basalts. Earth Planet Sci Lett 27:1–10

Leclerc AJ, Labeyrie LC (1987) Temperature dependence of oxygen isotopic fractionation between diatom silica and water. Earth Planet Sci Lett 84:69–74

Lehmann M, Siegenthaler U (1991) Equilibrium oxygen- and hydrogen-isotope fractionation between ice and water. J Glaciol 37:23–26

Letolle R (1980) Nitrogen-15 in the natural environment. In: Fritz P, Fontes JC (eds) Handbook of environmental isotope geochemistry. Elsevier, Amsterdam, pp 407–433

Leuenberger M, Siegenthaler U, Langway CC (1992) Carbon isotope composition of atmospheric CO_2 during the last ice age from an Antarctic ice core. Nature 357:488–490

Lewan MD (1983) Effects of thermal maturation on stable carbon isotopes as determined by hydrous pyrolysis of Woodford shale. Geochim Cosmochim Acta 47:1471–1480

Lewis RS, Anders E, Wright IP, Norris SJ, Pillinger CT (1983) Isotopically anomalous nitrogen in primitive meteorites. Nature 305:767–771

Lloyd MR (1967) Oxygen-18 composition of oceanic sulfate. Science 156:1228–1231

Lloyd MR (1968) Oxygen isotope behavior in the sulfate-water system. J Geophys Res 73:6099–6110

Logan GA, Hayes JM, Hieshima GB, Summons RE (1995) Terminal Proterozoic reorganization of biogeochemical cycles. Nature 376:53–56

Long A, Eastoe CJ, Kaufmann RS, Martin JG, Wirt L, Fincey JB (1993) High precision measurement of chlorine stable isotope ratios. Geochim Cosmochim Acta 57:2907–2912

Longinelli A (1984) Oxygen isotopes in mammal bone phosphate: a new tool for paleohydrological and paleoclimatological research? Geochim Cosmochim Acta 48:385–390

Longinelli A, Bartelloni M (1978) Atmospheric pollution in Venice, Italy, as indicated by isotopic analyses. Water Air Soil Poll 10:335–341

Longinelli A, Craig H (1967) Oxygen-18 variations in sulfate ions in sea-water and saline lakes. Science 156:56–59

Longinelli A, Edmond JM (1983) Isotope geochemistry of the Amazon basin. A reconnaissance. J Geophys Res 88:3703–3717

Longinelli A, Nuti S (1973) Revised phosphate-water isotopic temperature scale. Earth Planet Sci Lett 19:373–376

Longstaffe FJ (1989) Stable isotopes as tracers in clastic diagenesis. In: Hutcheon IE (ed) Short course in burial diagenesis. Min Assoc Canada Short Course Ser 15:201–277

Longstaffe FJ, Schwarcz HP (1977) $^{18}O/^{16}O$ of Archean clastic metasedimentary rocks: a petrogenetic indicator for Archean gneisses? Geochim Cosmochim Acta 41:1303–1312

Lorius C, Jouzel J, Ritz C, Merlivat L, Barkov NI, Korotkevich YS, Kotlyakov VM (1985) A 150,000 year climatic record from Antarctic ice. Nature 316:591–596

Luz B, Kolodny Y (1985) Oxygen isotope variations in phosphate of biogenic apatites, IV: mammal teeth and bones. Earth Planet Sci Lett 75:29–36

Machel HG, Krouse HR, Sassen R (1995) Products and distinguishing criteria of bacterial and thermochemical sulfate reduction. Appl Geochem 10:373–389

MacNamara J, Thode HG (1950) Comparison of the isotopic constitution of terrestrial and meteoritic sulphur. Phys Rev 78:307

Magaritz M (1991) Carbon isotopes, time boundaries and evolution. Terra Nova 3:251–256

Magenheim AJ, Spivack AJ, Volpe C, Ranson B (1994) Precise determination of stable chlorine isotope ratios in low-concentration natural samples. Geochim Cosmochim Acta 58:3117–3121

Magenheim AJ, Spivack AJ, Michael PJ, Gieskes JM (1995) Chlorine stable isotope composition of the oceanic crust: implications for earth's distribution of chlorine. Earth Planet Sci Lett 131:427–432

Mariotti A, Germon JC, Hubert P, Kaiser P, Letolle R, Tardieux P (1981) Experimental determination of nitrogen kinetic isotope fractionation: some principles, illustration for the denitrification and nitrification processes. Plant Soil 62: 413–430

Mariotti A, Germon JC, Leclerc A, Catroux G, Letolle R (1982) Experimental determination of kinetic isotope fractionation of nitrogen isotopes during denitrification. In: Schmidt HL, Förstel H, Heinzinger K (eds) Stable isotopes. Elsevier, New York

Marowsky G (1969) Schwefel-, Kohlenstoff- und Sauerstoffisotopenuntersuchungen am Kupferschiefer als Beitrag zur genetischen Deutung. Contrib Mineral Petrol 22:290–334

Marshall JD (1992) Climatic and oceanographic isotopic signals from the carbonate rock record and their preservation. Geol Mag 129:143–160

Matheney RK, Knauth LP (1989) Oxygen isotope fractionation between marine biogenic silica and seawater. Geochim Cosmochim Acta 53:3207–3214

Matsubaya O, Sakai H (1973) Oxygen and hydrogen isotopic study on the water of crystallization of gypsum from the Kuroko-type mineralization. Geochem J 7:153–165

Matsuhisa Y (1979) Oxygen isotopic compositions of volcanic rocks from the east Japan island arcs and their bearing on petrogenesis. J Volcanic Geotherm Res 5:271–296

Matsuhisa Y, Goldsmith JR, Clayton RN (1979) Oxygen isotope fractionation in the systems quartz-albite-anorthite-water. Geochim Cosmochim Acta 43:1131–1140

Matsumoto R (1992) Causes of the oxygen isotopic depletion of interstitial waters from sites 798 and 799, Japan Sea, Leg 128. Proc Ocean Drilling Prog Sci Res 127/128:697–703

Matsuo S, Friedman I, Smith GI (1972) Studies of Quaternary saline lakes. I. Hydrogen isotope fractionation in saline minerals. Geochim Cosmochim Acta 36:427–435

Mattey DP (1987) Carbon isotopes in the mantle. Terra Cognita 7:31–37

Mattey DP, Carr RH, Wright IP, Pillinger CT (1984) Carbon isotopes in submarine basalts. Earth Planet Sci Lett 70:196–206

Mattey DP, Lowry D, MacPherson C (1994) Oxygen isotope composition of mantle peridotites. Earth Planet Sci Lett 128:231–241

Matthews A, Goldsmith JR, Clayton RN (1983a) Oxygen isotope fractionation involving pyroxenes: the calibration of mineral-pair geothermometers. Geochim Cosmochim Acta 47:631–644

Matthews A, Goldsmith JR, Clayton RN (1983b) Oxygen isotope fractionation between zoisite and water. Geochim Cosmochim Acta 47:645–654

Matthews A, Goldsmith JR, Clayton RN (1983c) On the mechanics and kinetics of oxygen isotope exchange in quartz and feldspars at elevated temperatures and pressures. Geol Soc Am Bull 94:396–412

Mauersberger K (1981) Measurement of heavy ozone in the stratosphere. Geophys Res Lett 8:935–937

Mauersberger K (1987) Ozone isotope measurements in the stratosphere. Geophys Res Letter 14:80–83

McCaig AM, Wickham SM, Taylor HP (1990) Deep fluid circulation in Alpine shear zones, Pyrenees, France: field and oxygen isotope studies. Contr Mineral Petrol 106:41–60

McConnaughey T (1989a) ^{13}C and ^{18}O disequilibrium in biological carbonates.I. Patterns. Geochim Cosmochim Acta 53:151–162

McConnaughey T (1989b) ^{13}C and ^{18}O disequilibrium in biological carbonates. II. In vitro simulation of kinetic isotope effects. Geochim Cosmochim Acta 53:163–171

McCorkle DC, Emerson SR (1988) The relationship between pore water isotopic composition and bottom water oxygen concentration. Geochim Cosmochim Acta 52:1169–1178

McCorkle DC, Emerson SR, Quay P (1985) Carbon isotopes in marine porewaters. Earth Planet Sci Lett 74:13–26

McCrea JM (1950) The isotopic chemistry of carbonates and a paleotemperature scale. J Chem Phys 18:849–857

McCready RGL (1975) Sulphur isotope fractionation by Desulfovibrio and Desulfotomaculum species. Geochim Cosmochim Acta 39:1395–1401

McCready RGL, Kaplan IR, Din GA (1974) Fractionation of sulfur isotopes by the yeast Saccharomyces cerevisiae. Geochim Cosmochim Acta 38:1239–1253

McGregor ID, Manton SR (1986) Roberts Victor eclogites: ancient oceanic crust. J Geophys Res 91:14063–14079

McKeegan KD (1987) Ion microprobe measurements of H,C, O, Mg, and Si isotopic abundances in individual interplenary dust particles. PhD Thesis, Washington University, St Louis, Missouri

McKeegan KD, Walker RM, Zinner E (1985) Ion microprobe isotopic measurements of individual interplanetary dust particles. Geochim Cosmochim Acta 49:1971–1987

McKenzie J (1984) Holocene dolomitization of calcium carbonate sediments from the coastal sabkhas of Abu Dhabi, UAE: A stable isotope study. J Geol 89:185–198

McKibben MA, Eldridge CS (1995) Microscopic sulfur isotope variations in ore minerals from the Viburnum Trend, Southeast Missouri: a Shrimp study. Econ Geol 90:228–245

McKinney CR, McCrea JM, Epstein S, Allen HA, Urey HC (1950) Improvements in mass spectrometers for the measurement of small differences in isotope abundance ratios. Rev Sci Instrum 21:724

McMullen CC, Cragg CG, Thode HG (1961) Absolute ratio of $^{11}B/^{10}B$ in Searles Lake borax. Geochim Cosmochim Acta 23:147

Mekhtiyeva VL, Pankina GR (1968) Isotopic composition of sulfur in aquatic plants and dissolved sulfates. Geochemistry 5:624

Mekhtiyeva VL, Pankina GR, Gavrilov EY (1976) Distribution and isotopic composition of forms of sulfur in water animals and plants. Geochem Int 13:82

Melander L (1960) Isotope effects on reaction rates. Ronald, New York

Melander L, Saunders WH (1980) Reaction rates of isotopic molecules. Wiley, New York

Mengel K, Hoefs J (1990) Li – $\delta^{18}O$ – SiO_2 systematics in volcanic rocks and mafic lower crustal xenoliths. Earth Planet Sci Lett 101:42–53

Meyers PA (1994) Preservation of elemental and isotope source identification of sedimentary organic matter. Chem Geol 114:289–302

Ming T, Anders E, Hoppe P, Zinner E (1989) Meteoritic silicon carbide and its stellar sources, implications for galactic chemical evolution. Nature 339:351–354

Minigawa M, Wada E (1984) Stepwise enrichments of ^{15}N along food chains: further evidence and the relation between $\delta^{15}N$ and animal age. Geochim Cosmochim Acta 48:1135–1140

Moldovanyi EP, Lohmann KC (1984) Isotopic and petrographic record of phreatic diagenesis: Lower Cretaceous Sligo and Cupido Formations. J Sediment Petrol 54:972–985

Monson KD, Hayes JM (1982) Carbon isotopic fractionation in the biosynthesis of bacterial fatty acids. Ozonolysis of unsaturated fatty acids as a means of determining the intramolecular distribution of carbon isotopes. Geochim Cosmochim Acta 46:139–149

Monster J, Anders E, Thode HG (1965) $^{34}S/^{32}S$ ratios for the different forms of sulphur in the Orgueil meteorite and their mode of formation. Geochim Cosmochim Acta 29:773–779

Monster J, Appel PW, Thode HG, Schidlowski M, Carmichael CW, Bridgwater D (1979) Sulphur isotope studies in early Archean sediments from Isua, West Greenland: implications for the antiquity of bacterial sulfate reduction. Geochim Cosmochim Acta 43:405–413

Mook WG, Bommerson JC, Staverman WH (1974) Carbon isotope fractionation between dissolved bicarbonate and gaseous carbon dioxide. Earth Planet Sci Lett 22:169–176

Mook WG, Koopman M, Carter AF, Keeling CD (1983) Seasonal, latitudinal and secular variations in the abundance and isotopic ratios of atmospheric carbon dioxide. I. Results from land stations. J Geophys Res 88:10915–10933

Montoya JP, Horrigan SG, McCarthy JJ (1991) Rapid, storm-induced changes in the natural abundance of ^{15}N in a planktonic ecosystem, Chesapeake Bay, USA. Geochim Cosmochim Acta 55:3627–3638

Mossmann JR, Aplin AC, Curtis CD, Coleman ML (1991) Geochemistry of inorganic and organic sulfur in organic-rich sediments from the Peru Margin. Geochim Cosmochim Acta 55:3581–3595

Muehlenbachs K, Byerly G (1982) ^{18}O enrichment of silicic magmas caused by crystal fractionation at the Galapagos Spreading Center. Contr Mineral Petrol 79:76–79

Muehlenbachs K, Clayton RN (1972) Oxygen isotope studies of fresh and weathered submarine basalts. Can J Earth Sci 9:471–479

Muehlenbachs K, Clayton RN (1976) Oxygen isotope composition of the oceanic crust and its bearing on seawater. J Geophys Res 81:4365–4369

Nabelek PI (1991) Stable isotope monitors. In: Contact metamorphism. Rev Mineral 26:395–435

Nabelek PI, O'Neil JR, Papike JJ (1983) Vapor phase exsolution as a controlling factor in hydrogen isotope variation in granitic rocks: the Notch Peak granitic stock, Utah. Earth Planet Sci Lett 66:137–150

Nabelek PI, Labotka TC, O'Neil JR, Papike JJ (1984) Contrasting fluid/rock interaction between the Notch Peak granitic intrusion and argillites and limestones in western Utah: evidence from stable isotopes and phase assemblages. Contr Mineral Petrol 86:25–43

Nielsen H (1972) Sulphur isotopes and the formation of evaporite deposits. In: Geology of saline deposits, Proc Hannover Symp 1968, Earth Sci 7:91, UNESCO 1972

Nielsen H (1978) Sulfur isotopes. In: Wedepohl KH (ed) Handbook of geochemistry. Springer, Berlin Heidelberg New York

Nielsen H (1979) Sulfur isotopes. In: Jager E, Hunziker J (eds) Lectures in isotope geology. Springer, Berlin Heidelberg New York, pp 283–312

Nielsen H (1985) Isotope in der Lagerstättenforschung. In: Bender F (ed) Angewandte Geowissenschaften. Enke, Stuttgart

Nielsen H, Ricke W (1964) S-Isotopenverhaltnisse von Evaporiten aus Deutschland. Ein Beitrag zur Kenntnis von $\delta^{34}S$ im Meerwasser Sulfat. Geochim Cosmochim Acta 28:577–591

Nier AO (1950) A redetermination of the relative abundances of the isotopes of carbon, nitrogen, oxygen, argon and potassium. Phys Rev 77:789

Nier AO, Ney EP, Inghram MG (1947) A null method for the comparison of two ion currents in a mass spectrometer. Rev Sci Instrum 18:294

Nier AO, McElroy MB, Yung YL (1976) Isotopic composition of the Martian atmosphere. Science 194:68–70

Nissenbaum A, Presley BJ, Kaplan IR (1972) Early diagenesis in a reducing Fjord, Saanich Inlet, British Columbia. I: Chemical and isotopic changes in major components of interstitial water. Geochim Cosmochim Acta 36:1007–1027

Nitzsche HM, Stiehl G (1984) Untersuchungen zur Isotopenfraktionierung des Stickstoffs in den Systemen Ammonium/Ammoniak und Nitrid/Stickstoff. ZFI Mitt 84:283–291

Northrop DA, Clayton RN (1966) Oxygen isotope fractionations in systems containing dolomite. J Geol 74:174–196

Norton D, Taylor HP (1979) Quantitative simulation of the hydrothermal systems of crystallizing magmas on the basis of transport theory and oxygen isotope data: an analysis of the Skaergaard intrusion. J Petrol 20:421–486

Nriagu JO, Coker RD (1978) Isotopic composition of sulfur in precipitation within the Great Lakes Basin. Tellus 30:365–375

Nriagu JO, Coker RD, Barrie LA (1991) Origin of sulphur in Canadian Arctic haze from isotope measurements. Nature 349:142–145

Ohmoto H (1972) Systematics of sulfur and carbon isotopes in hydrothermal ore deposits. Econ Geol 67:551–578

Ohmoto H (1986) Stable isotope geochemistry of ore deposits. In: Rev Mineral 16:491–559

Ohmoto H, Rye RO (1979) Isotopes of sulfur and carbon. In: Geochemistry of hydrothermal ore deposits, 2nd edn. Rinehart and Winston, New York

Ohmoto H, Skinner BJ (1983) The Kuroko and related volcanogenic massive sulfide deposits. Econ Geol Monogr 5:•••••

Ohmoto H, Mizukani M, Drummond SE, Eldridge CS, Pisutha-Arnond V, Lenagh TC (1983) Chemical processes of Kuroko formation. Econ Geol Monogr 5:570–604

Ohmoto H, Kegawa T, Lowe DR (1993) 3.4 billion year old biogenic pyrites from Barberton, South Africa: sulfur isotope evidence. Science 262:555

O'Leary MH (1981) Carbon isotope fractionation in plants. Phytochemistry 20:553–567

Oliver NHS, Hoering TC, Johnson TW, Rumble D, Shanks WC (1992) Sulfur isotope disequilibrium and fluid/rock interaction during metamorphism of sulfidic black shales from the Waterville-Augusta area, Maine, USA. Geochim Cosmochim Acta 56:4257–4265

O'Neil JR (1986) Theoretical and experimental aspects of isotopic fractionation. In: Stable isotopes in high temperature geological processes. Rev Mineral 16:1–40

O'Neil JR, Truesdell AH (1991) Oxygen isotope fractionation studies of solute-water interactions. In: Stable isotope geochemistry: a tribute to Samuel Epstein. Geochem Soc Spec Publ 3:17–25

O'Neil JR, Roe LJ, Reinhard E, Blake RE (1994) A rapid and precise method of oxygen isotope analysis of biogenic phosphate. Isr J Earth Sci 43:203–212

Ongley JS, Basu AR, Kyser TK (1987) Oxygen isotopes in coexisting garnets, clinopyroxenes and phlogopites of Roberts Victor eclogites: implications for petrogenesis and mantle metasomatism. Earth Planet Sci Lett 83:80–84

Onuma N, Clayton RN, Mayeda TK (1970) Oxygen isotope fractionation between minerals and an estimate of the temperature of formation. Science 167:536–538

Ott U (1993) Interstellar grains in meteorites. Nature 364:25–33

Owen T, Biemann K, Rushneck DR, Biller JE, Howarth DW, Lafleur AL (1977) The composition of the atmosphere at the surface of Mars. J Geophys Res 82:4635–4639

Owen T, Maillard JP, DeBergh C, Lutz BL (1988) Deuterium on Mars: the abundance of HDO and the value of D/H. Science 240:1767–1770

Owens NJP (1987) Natural variations in ^{15}N in the marine environment. Adv Mar Biol 24:390–451

Palmer MR (1991) Boron isotope systematics of Halmahera (Indonesia) arc lavas: evidence for involvement of the subducted slab. Geology 19:215–217

Palmer MR, Slack JF (1989) Boron isotopic composition of tourmaline from massive sulfide deposits and tourmalinites. Contr Mineral Petrol 103:434–451

Palmer MR, Spivack AJ, Edmond JM (1987) Temperature and pH controls over isotopic fractionation during the adsorption of boron on marine clay. Geochim Cosmochim Acta 51:2319–2323

Park R, Epstein S (1960) Carbon isotope fractionation during photosynthesis. Geochim Cosmochim Acta 21:110–126

Parker PL (1964) The biogeochemistry of the stable isotopes of carbon in a marine bay. Geochim Cosmochim Acta 28 (1):155–1164

Pawellek F, Veizer J (1994) Carbon cycle in the upper Danube and its tributaries: $\delta^{13}C_{DIC}$ constraints. Isr J Earth Sci 43:187–194

Perry EA, Gieskes JM, Lawrence JR (1976) Mg, Ca and $^{18}O/^{16}O$ exchange in the sediment-pore water system, Hole 149, DSDP. Geochim Cosmochim Acta 40:413–423

Peters MT, Wickham SM (1995) On the causes of ^{18}O depletion and $^{18}O/^{16}O$ homogenization during regional metamorphism, the east Humboldt Range core complex, Nevada. Contr Mineral Petrol 119:68–82

Peters KE, Rohrbach BG, Kaplan IR (1981) Carbon and hydrogen stable isotope variations in kerogen during laboratory-simulated thermal maturation. Am Assoc Petrol Geol Bull 65:501–508

Peterson BJ, Fry B (1987) Stable isotopes in ecosystem studies. Annu Rev Ecol Syst 18:293–320

Phillips FM, Bentley HW (1987) Isotopic fractionation during ion filtration: I. Theory. Geochim Cosmochim Acta 51: 683–695

Pillinger CT (1984) Light element stable isotopes in meteorites – from grams to picograms. Geochim Cosmochim Acta 48: 2739–2768

Pineau F, Javoy M (1983) Carbon isotopes and concentrations in mid-ocean ridge basalts. Earth Planet Sci Lett 62:239–257

Pineau F, Javoy M, Bottinga Y (1976) $^{13}C/^{12}C$ ratios of rocks and inclusions in popping rocks of the Mid-Atlantic Ridge and their bearing on the problem of isotopic composition of deep-seated carbon. Earth Planet Sci Lett 29:413–421

Polyakov VB, Kharlashina NN (1994) Effect of pressure on equilibrium isotope fractionation. Geochim Cosmochim Acta 58:4739–4750

Poorter RPE, Varekamp JC, Poreda RJ, Van Bergen MJ, Kreulen R (1991) Chemical and isotopic compositions of volcanic gases from the east Sunda and Banda arcs, Indonesia. Geochim Cosmochim Acta 55:3795–3807

Popp B, Takigiku R, Hayes JM, Louda JW, Baker EW (1989) The post Paleozoic chronology and mechanism of ^{13}C depletion in primary organic matter. Am J Sci 289:436–454

Poreda R (1985) Helium-3 and deuterium in back arc basalts: Lau Basin and the Mariana trough. Earth Planet Sci Lett 73:244–254

Poreda R, Schilling JG, Craig H (1986) Helium and hydrogen isotopes in ocean-ridge basalts north and south of Iceland. Earth Planet Sci Lett 78:1–17

Poreda RJ, Craig H, Arnorsson S, Welhan JA (1992) Helium isotopes in Icelandic geothermal systems. I. ^{3}He, gas chemistry and ^{13}C relations. Geochim Cosmochim Acta 56:4221–4228

Price FT, Shieh YN (1979) The distribution and isotopic composition of sulfur in coals from the Illinois Basin. Econ Geol 74:1445–1461

Prombo CA, Clayton RN (1985) A striking nitrogen isotope anomaly in the Bencubbin and Weatherford meteorites. Science 230:935–937

Puchelt H, Sabels BR, Hoering TC (1971) Preparation of sulfur hexafluoride for isotope geochemical analysis. Geochim Cosmochim Acta 35:625–628

Quay PD et al. (1991) Carbon isotopic composition of CH_4: fossil and biomass burning source strengths. Glob Biogeochem Cycles 5:25–47

Quay PD, Tilbrook B, Wong CS (1992) Oceanic uptake of fossil fuel CO_2: carbon-13 evidence. Science 256:74–79

Raab M, Spiro B (1991) Sulfur isotopic variations during seawater evaporation with fractional crystallization. Chem Geol 86:323–333

Rabinowitch EI (1945) Photosynthesis and related processes, vol I. Interscience, New York, p 10

Rabinovich AL, Grinenko VA (1979) Sulfate sulfur isotope ratios for USSR river water. Geochemistry 16 (2):68–79

Rafter TA (1957) Sulphur isotopic variations in nature, P 1: the preparation of sulphur dioxide for mass spectrometer examination. NZ J Sci Tech B38:849

Railsback LB (1990) Influence of changing deep circulation on the Phanerozoic oxygen isotopic record. Geochim Cosmochim Acta 54:1501–1509

Raiswell R, Berner RA (1985) Pyrite formation in euxinic and semi-euxinic sediments. Am J Sci 285:710–724

Rakestraw NM, Rudd DP, Dole M (1951) Isotopic composition of oxygen in air dissolved in Pacific Ocean water as a function of depth. J Am Chem Soc 73:2976

Ransom B, Spivack AJ, Kastner M (1995) Stable Cl isotopes in subduction-zone pore waters: implications for fluid-rock reactions and the cycling of chlorine. Geology 23:715–718

Rau GH, Sweeney RE, Kaplan IR (1982) Plankton $^{13}C/^{12}C$ ratio changes with latitude: differences between northern and southern oceans. Deep Sea Res 29:1035–1039

Rau GH, Arthur MA, Dean WE (1987) $^{15}N/^{14}N$ variations in Cretaceous Atlantic sedimentary sequences: implication for past changes in marine nitrogen biochemistry. Earth Planet Sci Lett 82:269–279

Rau GH, Takahashi T, DesMarais DJ (1989) Latitudinal variations in plankton ^{13}C: implications for CO_2 and productivity in past ocean. Nature 341:516–518

Rau GH, Takahashi T, DesMarais DJ, Repeta DJ, Martin JH (1992) The relationship between $\delta^{13}C$ of organic matter and $\Sigma CO_{2(aq)}$ in ocean surface water: data from a JGOFS site in the northeast Atlantic Ocean and a model. Geochim Cosmochim Acta 56:1413–1419

Rayleigh JWS (1896) Theoretical considerations respecting the separation of gases by diffusion and similar processes. Philos Mag 42:493

Raymo ME, Ruddiman WF (1992) Tectonic forcing of late Cenozoic climate. Nature 359:117–1222

Redding CE, Schoell M, Monin JC, Durand B (1980) Hydrogen and carbon isotopic composition of coals and kerogen. Phys Chem Earth 12:711–723

Redfield AC, Friedman I (1965) Factors affecting the distribution of deuterium in the ocean. Symp Rhode Island Occ Publ 3:149

Rees CE (1978) Sulphur isotope measurements using SO_2 and SF_6. Geochim Cosmochim Acta 42:383–389

Rees CE, Jenkins WJ, Monster J (1978) The sulphur isotopic composition of ocean water sulphate. Geochim Cosmochim Acta 42:377–381

Rice DD, Claypool GE (1981) Generation, accumulation and resource potential of biogenic gas. Am Assoc Petrol Geol Bull 65:5–25

Richet P, Bottinga Y, Javoy M (1977) A review of H, C, N, O, S, and Cl stable isotope fractionation among gaseous molecules. Annu Rev Earth Planet Sci 5:65–110

Richter R, Hoernes S (1988) The application of the increment method in comparison with experimentally derived and calculated O-isotope fractionations. Chemie der Erde 48:1–18

Ricke W (1964) Präparation von Schwefeldioxid zur massenspektrometrischen Bestimmung des S-Isotopenverhältnisses in natürlichen S-Verbindungen. Z Anal Chemie 199:401

Rindsberger MS, Jaffe S, Rahamin S, Gat JR (1990) Patterns of the isotopic composition of precipitation in time and space; data from the Israeli storm water collection program. Tellus 42B:263–271

Robert F, Epstein S (1980) Carbon, hydrogen and nitrogen isotopic composition of the Renazzo and Orgeuil organic components. Meteoritics 15:351

Robert F, Epstein S (1982) The concentration and isotopic composition of hydrogen, carbon and nitrogen carbonaceous meteorites. Geochim Cosmochim Acta 46:81–95

Robert F, Merlivat L, Javoy M (1978) Water and deuterium content in ordinary chondrites. Meteoritics 12:349–354

Robert F, Merlivat L, Javoy M (1979a) Water and deuterium content in the Chainpur meteorite. Meteoritics 13:613–615

Robert F, Merlivat L, Javoy M (1979b) Deuterium concentration in the early solar system: a hydrogen and oxygen isotope study. Nature 282:785–789

Robinson BW, Kusakabe M (1975) Quantitative preparation of sulphur dioxide for $^{34}S/^{32}S$ analyses from sulphides by combustion with cuprous oxide. Anal Chem 47:1179

Rooney MA, Claypool GE, Chung HM (1995) Modeling thermogenic gas generation using carbon isotope ratios of natural gas hydrocarbons. Chem Geol 126:219–232

Rosenbaum J, Sheppard SMF (1986) An isotopic study of siderites, dolomites and ankerites at high temperatures. Geochim Cosmochim Acta 50:1147–1150

Rozanski K, Araguas-Araguas L, Gonfiantini R (1993) Isotopic patterns in modern global precipitation. In: Climate change in continental isotopic records. Geophys Monogr 78:1–36

Rubinson M, Clayton RN (1969) Carbon-13 fractionation between aragonite and calcite. Geochim Cosmochim Acta 33:997–1002

Rye RO (1974) A comparison of sphalerite-galena sulfur isotope temperatures with filling-temperatures of fluid inclusions. Econ Geol 69:26–32

Rye RO (1993) The evolution of magmatic fluids in the epithermal environment: the stable isotope perspective. Econ Geol 88:733–753

Rye RO, Ohmoto H (1974) Sulfur and carbon isotopes and ore genesis. A review. Econ Geol 69:826–842

Rye RO, Schuiling RD, Rye DM, Jansen JBH (1976) Carbon, hydrogen and oxygen isotope studies of the regional metamorphic complex at Naxos, Greece. Geochim Cosmochim Acta 40:1031–1049

Rye RO, Bethke PM, Wasserman MD (1992) The stable isotope geochemistry of acid sulfate. Econ Geol 87:227–262

Sackett WM (1988) Carbon and hydrogen isotope effects during the thermocatalytic production of hydrocarbons in laboratory simulation experiments. Geochim Cosmochim Acta 42:571–580

Sackett WM, Thompson RR (1963) Isotopic organic carbon composition of recent continental derived clastic sediments of Eastern Gulf Coast, Gulf of Mexico. Bull Am Assoc Petrol Geol 47:525

Sackett WM, Eadie BJ, Exner ME (1973) Stable isotope composition of organic carbon in Recent Antarctic sediments. Adv Org Geochem 1973:661

Saino T, Hattori A (1980) ^{15}N natural abundance in oceanic suspended particulate organic matter. Nature 283:752–754

Saino T, Hattori A (1987) Geophysical variation of the water column distribution of suspended particulate organic nitrogen and its ^{15}N natural abundance in the Pacific and its marginal seas. Deep Sea Res 34:807–827

Sakai H (1957) Fractionation of sulphur isotopes in nature. Geochim Cosmochim Acta 12:150–169

Sakai H (1968) Isotopic properties of sulfur compounds in hydrothermal processes. Geochem J 2:29–49

Sakai H, Matsubaya O (1974) Isotopic geochemistry of the thermal waters of Japan and its bearing on the Kuroko ore solutions. Econ Geol 69:974–991

Sakai H, Casadevall TJ, Moore JG (1982) Chemistry and isotope ratios of sulfur in basalts and volcanic gases at Kilauea volcano, Hawaii. Geochim Cosmochim Acta 46:729–738

Sakai H, DesMarais DJ, Ueda A, Moore JG (1984) Concentrations and isotope ratios of carbon, nitrogen and sulfur in ocean-floor basalts. Geochim Cosmochim Acta 48:2433–2441

Salomons W, Mook WG (1981) Field observations of the isotope composition of particulate organic carbon in the Southern North Sea and adjacent estuaries. Mar Geol 41:M11–M20

Sanyal A, Hemming NG, Hanson GN, Broecker WS (1995) Evidence for a higher pH in the glacial ocean from boron isotopes in foraminifera. Nature 373:234–236

Sarntheim M, Winn K, Jung SJA, Duplessy JC, Labeyrie L, Erlenkeuser H, Ganssen G (1994) Changes in east Atlantic deepwater circulation over the last 30|000 years: eight time slice reconstructions. Paleoceanography 9:209–268

Sasaki A, Arikawa Y, Folinsbee RE (1979) Kiba reagent method of sulfur extraction applied to isotopic work. Bull Geol Surv Jpn 30:241

Sass E, Kolodny Y (1972) Stable isotopes, chemistry and petrology of carbonate concretions (Mishash formation, Israel). Chem Geol 10:261–286

Savin SM, Epstein S (1970a) The oxygen and hydrogen isotope geochemistry of clay minerals. Geochim Cosmochim Acta 34:25–42

Savin SM, Epstein S (1970b) The oxygen and hydrogen isotope geochemistry of ocean sediments and shales. Geochim Cosmochim Acta 34:43–63

Savin SM, Epstein S (1970c) The oxygen isotope composition of coarse grained sedimentary rocks and minerals. Geochim Cosmochim Acta 34:323–329

Savin SM, Lee M (1988) Isotopic studies of phyllosilicates. Rev Mineral 19:189–223

Scheele N, Hoefs J (1992) Carbon isotope fractionation between calcite, graphite and CO_2. Contr Mineral Petrol 112:35–45

Schidlowski M, Hayes JM, Kaplan IR (1983) Isotopic inferences of ancient biochemistries: carbon, sulfur, hydrogen and nitrogen. In: Schopf JW (ed) Earth's earliest biosphere: its origin and evolution. Princeton University Press, Princeton, pp 149–186

Schiegl WE, Vogel JV (1970) Deuterium content of organic matter. Earth Planet Sci Lett 7:307–313

Schirmer T (1995) Die Zusammensetzung (D, ^{18}O und ausgewählte Inhaltsstoffe) von Einzelniederschlagsereignissen Göttingens und Clausthal-Zellerfelds für den Zeitraum vom Mai 93 bis zum März 94. Diplomarbeit, University of Göttingen

Schmidt M, Botz R, Stoffers P, Anders T, Bohrmann G (1996) Oxygen isotopes in marine diatoms: a comparative study of analytical techniques and new results on the isotope fractionation during phytoplankton growth. Geochim Cosmochim Acta (submitted)

Schoell M (1980) The hydrogen and carbon isotopic composition of methane from natural gases of various origins. Geochim Cosmochim Acta 44:649–661

Schoell M (1983) Genetic characterization of natural gases. Bull Am Assoc Petrol Geol 67:2225–2238

Schoell M (1984a) Recent advances in petroleum isotope geochemistry. Organ Geochem 6:645–663

Schoell M (1984b) Wasserstoff-und Kohlenstoffisotope in organischen Substanzen, Erdölen und Erdgasen. Geol Jahrb R D, H 67

Schoell M (1988) Multiple origins of methane in the earth. Chem Geol 71:1–10

Schoell M, McCaffrey MA, Fago FJ, Moldovan JM (1992) Carbon isotope compositions of 28,30-bisnorhopanes and other biological markers in a Monterey crude oil. Geochim Cosmochim Acta 56:1391–1399

Schoeller DA, Peterson DW, Hayes JM (1983) Double-comparison method for mass spectrometric determination of hydrogen isotopic abundances. Anal Chem 55:827–832

Schoenheimer R, Rittenberg D (1939) Studies in protein metabolism: I. General considerations in the application of isotopes to the study of protein metabolism. The normal abundance of nitrogen isotopes in amino acids. J Biol Chem 127:285–290

Schoeninger MJ, DeNiro MJ (1984) Nitrogen and carbon isotopic composition of bone collagen from marine and terrestrial animals. Geochim Cosmochim Acta 48:625–639

Scholten SO (1991) The distribution of nitrogen isotopes in sediments. PhD Thesis, University of Utrecht

Schütze H (1980) Der Isotopenindex – eine Inkrementmethode zur näherungsweisen Berechnung von Isotopenaustauschgleichgewichten zwischen kristallinen Substanzen. Chemie Erde 39:321–334

Schwarcz HP, Cortecci G (1974) Isotopic analyses of spring and stream water sulfate from the Italian Alps and Appennines. Chem Geol 13: 285–294

Schwarcz HP, Agyei EK, McCullen CC (1969) Boron isotopic fractionation during clay adsorption from seawater. Earth Planet Sci Lett 6:1–5

Schwarcz, Melbye J, Katzenberg MA, Knyf M (1985) Stable isotopes in human skeletons of southern Ontario: reconstruction of palaeodiet. J Archaeol Sci 12:187–206

Seccombe PK, Spry PG, Both Ra, Jones MT, Schiller JC (1985) Base metal mineralization in the Kaumantoo Group, South Australia: a regional sulfur isotope study. Econ Geol 80:1824–1841

Shackleton NJ, Kennett JP (1975) Paleotemperature history of the Cenozoic and initiation of Antarctic glaciation: oxygen and carbon isotope analyses in DSDP sites 277, 279 and 281. Initial Rep DSDP 29:743–755

Shemesh A, Burckle LH, Hays JD (1995) Late Pleistocene oxygen isotope records of biogenic silica from the Atlantic sector of the southern ocean. Paleoceanography 10:179–196

Shackleton NJ, Hall MA, Line J, Cang S (1983) Carbon isotope data in core V19-30 confirm reduced carbon dioxide concentration in the ice age atmosphere. Nature 306:319–322

Sharp ZD (1990) A laser-based microanalytical method for the in situ determination of oxygen isotope ratios of silicates and oxides. Geochim Cosmochim Acta 54:1353–1357

Sharp ZD (1995) Oxygen isotope geochemistry of the Al_2SiO_5 polymorphs. Am J Sci 295:1058–1076

Shelton KL, Rye DM (1982) Sulfur isotopic compositions of ores from Mines Gaspe, Quebec: an example of sulfate-sulfide isotopic disequilibria in ore forming fluids with applications to other porphyry type deposits. Econ Geol 77:1688–1709

Shemesh A, Kolodny Y, Luz B (1983) Oxygen isotope variations in phosphate of biogenic apatites, II. Phosphorite rocks. Earth Planet Sci Lett 64: 405–416

Shemesh A, Kolodny Y, Luz B (1988) Isotope geochemistry of oxygen and carbon in phosphate and carbonate of phosphorite francolite. Geochim Cosmochim Acta 52:2565–2572

Shemesh A, Charles CD, Fairbanks RG (1992) Oxygen isotopes in biogenic silica: global changes in ocean temperature and isotopic composition. Science 256:1434–1436

Sheppard SMF (1986) Characterization and isotopic variations in natural waters. In: Stable isotopes in high temperature geological processes. Rev Mineral 16:165–183

Sheppard SMF, Epstein S (1970) D/H and O^{18}/O^{16} ratios of minerals of possible mantle or lower crustal origin. Earth Planet Sci Lett 9:232–239

Sheppard SMF, Gilg HA (1995) Stable isotope geochemistry of clay minerals. Clay Miner 31:1–24

Sheppard SMF, Harris C (1985) Hydrogen and oxygen isotope geochemistry of Ascension Island lavas and granites: variation with crystal fractionation and interaction with sea water. Contrib Mineral Petrol 91:74–81

Sheppard SMF, Schwarcz HP (1970) Fractionation of carbon and oxygen isotopes and magnesium between coexisting metamorphic calcite and dolomite. Contr Mineral Petrol 26:161–198

Sheppard SMF, Nielsen RL, Taylor HP (1969) Oxygen and hydrogen isotope ratios of clay minerals from Porphyry Copper Deposits. Econ Geol 64:755–777

Sheppard SMF, Nielsen RL, Taylor HP (1971) Hydrogen and oxygen isotope ratios in minerals from Porphyry Copper Deposits. Econ Geol 66:515–542

Shieh YN, Schwarcz HP (1974) Oxygen isotope studies of granite and migmatite, Grenville province of Ontario, Canada. Geochim Cosmochim Acta 38:21–45

Simon K (1996) Does δD from fluid inclusion in quartz reflect the original hydrothermal fluid? Geochim Cosmochim Acta (submitted)

Skauli H, Boyle AJ, Fallick AE (1992) A sulphur isotope study of the Bleikvassli Zn-Pb-Cu deposit, Nordland, northern Norway. Miner Depos 27:284–292

Skirrow R, Coleman ML (1982) Origin of sulfur and geothermometry of hydrothermal sulfides from the Galapagos Rift, 86 W. Nature 249:142–144

Smith BN, Epstein S (1970) Biochemistry of the stable isotopes of hydrogen and carbon in salt marsh biota. Plant Physiol 46:738

Smith BN, Epstein S (1971) Two categories of $^{13}C/^{12}C$ ratios for higher plants. Plant Physiol 47:380

Smith JW, Batts BD (1974) The distribution and isotopic composition of sulfur in coal. Geochim Cosmochim Acta 38: 121–123

Smith JW, Gould KW, Rigby D (1982) The stable isotope geochemistry of Australian coals. Org Geochem 3:111–131

Smith JW, Rigby D, Schmidt PW, Clark DA (1983) D/H ratios of coals and the palaeolatitude of their deposition. Nature 302:322–323

Sofer Z (1978) Isotopic composition of hydration water in gypsum. Geochim Cosmochim Acta 42:1141–1149

Sofer Z (1984) Stable carbon isotope compositions of crude oils: application to source deposi-
    tional environments and petroleum alteration. Am Assoc Petrol Geol Bull 68:31–49
Sofer Z, Gat JR (1972) Activities and concentrations of oxygen-18 in concentrated aqueous salt so-
    lutions: analytical and geophysical implications. Earth Planet Sci Lett 15:232–238
Sofer Z, Gat JR (1975) The isotopic composition of evaporating brines: effect of the isotopic ac-
    tivity ratio in saline solutions. Earth Planet Sci Lett 26:179–186
Sowers T, Bender M, Raynaud D, Korotkevich YS, Orchardo J (1991) The $\delta^{18}O$ of atmospheric O_2
    from air inclusions in the Vostok ice core: timing of CO_2 and ice volume changes during the
    Penultimate deglaciation. Paleoceanography 6:679–696
Spindel W, Stern MJ, Monse EU (1970) Further studies on temperature dependences of isotope ef-
    fects. J Chem Phys 2:2022–2033
Spivack AJ, Edmond JM (1986) Determination of boron isotope ratios by thermal ionization mass
    spectrometry of the dicesium metaborate cation. Anal Chem 58:31–35
Spivack AJ, Edmond JM (1987) Boron isotope exchange between seawater and the oceanic crust.
    Geochim Cosmochim Acta 51:1033–1043
Spivack AJ, You CF, Smith J (1993) Foraminiferal boron isotope ratios as a proxy for surface ocean
    pH over the past 21 Myr. Nature 363:149–151
Stahl W (1977) Carbon and nitrogen isotopes in hydrocarbon research and exploration. Chem
    Geol 20:121–149
Stahl W, Carey BD (1975) Source-rock identification by isotope analyses of natural gases from
    fields in the Val Verde and Delaware Basins, West Texas. Chem Geol 16:257–267
Stern MJ, Spindel W, Monse EU (1968) Temperature dependence of isotope effects. J Chem Phys
    48:2908
Stevens LM (1988) Atmospheric methane. Chem Geol 71:11–21
Stewart MK (1974) Hydrogen and oxygen isotope fractionation during crystallization of mirabi-
    lite and ice. Geochim Cosmochim Acta 38:167–172
Stolper EM (1982) Water in silicate glasses: an infrared spectroscopic study. Contr Mineral Petrol
    81:1–17
Stueber AM, Walter LM (1991) Origin and chemical evolution of formation waters from Silur-
    ian–Devonian strata in the Illinois basin. Geochim Cosmochim Acta 55:309–325
Styrt MM, Brackmann AJ, Holland HD, Clark BC, Pisutha-Arnold U, Eldridge CS, Ohmoto H
    (1981) The mineralogy and the isotopic composition of sulfur in hydrothermal sulfide/sulfate
    deposits on the East Pacific Rise, 21°N latitude. Earth Planet Sci Lett 53:382–390
Summons RE, Hayes JM (1992) Principles of molecular and isotopic biochemistry. In: Schopf JW,
    Klein C (eds) The Proterozoic biosphere: a multidisciplinary study. Oxford University Press,
    Oxford
Suzuoki T, Epstein S (1976) Hydrogen isotope fractionation between OH-bearing minerals and
    water. Geochim Cosmochim Acta 40:1229–1240
Swart PK, Burns SJ, Leder JJ (1991) Fractionation of the stable isotopes of oxygen and carbon in
    carbon dioxide during the reaction of calcite with phosphoric acid as a function of tempera-
    ture and technique. Chem Geol 86:89–96
Sweeney RE, Kaplan IR (1980) Natural abundance of ^{15}N as a source indicator for near-shore ma-
    rine sedimentary and dissolved nitrogen. Mar Chem 9:81–94
Sweeney RE, Liu KK, Kaplan IR (1978) Oceanic nitrogen isotopes and their use in determining
    the source of sedimentary nitrogen. DSIR Bull 220:9–26
Swihart GH, Moore PB (1989) A reconnaissance of the boron isotopic composition of tourmaline.
    Geochim Cosmochim Acta 53:911–916
Swihart GH, Moore PB, Callis EL (1986) Boron isotopic composition of marine and non-marine
    evaporite borates. Geochim Cosmochim Acta 50:1297–1301
Sywall M (1995) Variation der Lithium Isotopenzusammensetzung in der Natur: Konsequenzen
    für den globalen Lithium Kreislauf. Dissertation der Math-Nat Fachbereiche, Universität Göt-
    tingen
Talbot MR (1990) A review of the palaeohydrological interpretation of carbon and oxygen isoto-
    pic ratios in primary lacustrine carbonates. Chem Geol 80:261–279
Tan FC, Strain PM (1985) Sources, sinks and distribution of organic carbon in the St. Lawrence
    estuary, Canada. Geochim Cosmochim Acta 47:125–132
Tarutani T, Clayton RN, Mayeda TK (1969) The effect of polymorphism and magnesium substi-
    tution on oxygen isotope fractionation between calcium carbonate and water. Geochim Cos-
    mochim Acta 33:987–996
Taube H (1954) Use of oxygen isotope effects in the study of hydration ions. J Phys Chem 58:523
Taylor BE (1986) Magmatic volatiles: isotopic variation of C, H and S. Rev Mineral 16:185–225

Taylor BE (1987) Stable isotope geochemistry of ore-forming fluids. In: Stable isotope geochemistry of low-temperature fluids. Short Course Min Assoc Can 13:337–445

Taylor BE, O'Neil JR (1977) Stable isotope studies of metasomatic Ca-Fe-Al-Si skarns and associated metamorphic and igneous rocks, Osgood Mountains, Nevada. Contr Mineral Petrol 63:1–49

Taylor BE, Bucher-Nurminen K (1986) Oxygen and carbon isotope and cation geochemistry of metasomatic carbonates and fluids – Bergell aureole, Northern Italy. Geochim Cosmochim Acta 50:1267–1279

Taylor BE, Eichelberger JC, Westrich HR (1983) Hydrogen isotopic evidence of rhyolitic magma degassing during shallow intrusion and eruption. Nature 306:541–545

Taylor HP (1967) Oxygen isotope studies of hydrothermal mineral deposits. In: Geochemistry of hydrothermal ore deposits. Rinehart and Winston, New York

Taylor HP (1968) The oxygen isotope geochemistry of igneous rocks. Contr Mineral Petrol 19:1–71

Taylor HP (1974) The application of oxygen and hydrogen isotope studies to problems of hydrothermal alteration and ore deposition. Econ Geol 69:843–883

Taylor HP (1977) Water/rock interactions and the origin of H_2O in granite batholiths. J Geol Soc 133:509

Taylor HP (1978) Oxygen and hydrogen isotope studies of plutonic granitic rocks. Earth Planet Sci Lett 38:177–210

Taylor HP (1980) The effects of assimilation of country rocks by magmas on $^{18}O/^{16}O$ and $^{87}Sr/^{86}Sr$ systematics in igneous rocks. Earth Planet Sci Lett 47:243–254

Taylor HP (1986) Igneous rocks: II. Isotopic case studies of circumpacific magmatism. In: Stable isotopes in high temperature geological processes. Rev Mineral 16:273–317

Taylor HP (1987) Comparison of hydrothermal systems in layered gabbros and granites, and the origin of low-$\delta^{18}O$ magmas. In: Magmatic processes: physicochemical principles. Geochemical Soc Spec Publ 1:337–357

Taylor HP (1988) Oxygen, hydrogen and strontium isotope constraints on the origin of granites. Trans R Soc Edinb Earth Sci 79:317–338

Taylor HP, Epstein S (1962) Relation between $^{18}O/^{16}O$ ratios in coexisting minerals of igneous and metamorphic rocks. I. Principles and experimental results. Geol Soc Am Bull 73:461–480

Taylor HP, Forester RW (1979) An oxygen and hydrogen isotope study of the Skaergaard intrusion and its country rocks: a description of a 55 MY old fossil hydrothermal system. J Petrol 20:355–419

Taylor HP, Sheppard SMF (1986) Igneous rocks: I. Processes of isotopic fractionation and isotope systematics. In: Stable isotopes in high temperature geological processes. Rev Mineral 16:227–271

Taylor HP, Urey HC (1938) Fractionation of the lithium and potassium isotopes by chemical exchange with zeolites. J Chem Phys 6:429–438

Taylor HP, Turi B, Cundari A (1984) $^{18}O/^{16}O$ and chemical relationships in K-rich volcanic rocks from Australia, East Africa, Antarctica and San Venanzo Cupaello, Italy. Earth Planet Sci Lett 69:263–276

Thiemens MH (1988) Heterogeneity in the nebula: evidence from stable isotopes. In: Kerridge JF, Matthews MS (eds) Meteorites and the early solar system. University of Arizona Press, Arizona, pp 899–923

Thiemens MH, Heidenreich JE (1983) The mass independent fractionation of oxygen – a novel isotope effect and its cosmochemical implications. Science 219:1073–1075

Thiemens MH, Trogler WH (1991) Nylon production: an unknown source of atmospheric nitrous oxide. Science 251:932–934

Thiemens MH, Jackson T, Mauersberger K, Schueler B, Morton J (1991) Oxygen isotope fractionation in stratospheric CO_2. Geophys Res Lett 18:669–672

Thiemens MH, Jackson T, Zipf EC, Erdman PW, van Egmond C (1995) Carbon dioxide and oxygen isotope anomalies in the mesophere and stratosphere. Science 270:969–972

Thode HG, Monster J (1964) The sulfur isotope abundances in evaporites and in ancient oceans. In: Vinogradov AP (ed) Proceedings of the geochemical conference commemorating the centenary of V.I. Vernadskii's birth, vol 2, 630pp

Thode HG, Macnamara J, Collins CB (1949) Natural variations in the isotopic content of sulphur and their significance. Can J Res 27B:361

Todd CS, Evans BW (1993) Limited fluid-rock interaction at marble-gneiss contacts during Cretaceous granulite-facies metamorphism, Seward Peninsula, Alaska. Contr Mineral Petrol 114:27–41

Touret J (1971) Le facies granulite en Norvege meridionale. Les inclusions fluids. Lithos 4:423–436

Trofimov A (1949) Isotopic constitution of sulfur in meteorites and in terrestrial objects (in Russian). Dokl Akad Nauk SSSR 66:181

Trudinger PA, Chambers LA, Smith JW (1985) Low temperature sulphate reduction: biological versus abiological. Can J Earth Sci 22:1910–1918

Truesdell AH (1974) Oxygen isotope activities and concentrations in aqueous salt solution at elevated temperatures: consequences for isotope geochemistry. Earth Planet Sci Lett 23:387–396

Truesdell AH, Hulston JR (1980) Isotopic evidence on environments of geothermal systems. In: Fritz P, Fontes J (eds) Handbook of environmental isotope geochemistry, vol I. Elsevier, New York, pp 179–226

Tucker ME, Wright PV (1990) Carbonate sedimentology. Blackwell, London, pp 365–400

Tudge AP (1960) A method of analysis of oxygen isotopes in orthophosphate – its use in the measurement of paleotemperatures. Geochim Cosmochim Acta 18:81–93

Turner JV (1982) Kinetic fractionation of carbon-13 during calcium carbonate precipitation. Geochim Cosmochim Acta 46:1183–1192

Ueda A, Sakai H (1983) Simultaneous determinations of the concentration and isotope ratio of sulfate-and sulfide-sulfur and carbonate-carbon in geological samples. Geochemical J 17:185–196

Ueda A, Sakai S (1984) Sulfur isotope study of Quaternary volcanic rocks from the Japanese Island. Arc. Geochim Cosmochim Acta 48:1837–1848

Urey HC, Brickwedde FG, Murphy GM (1932a) A hydrogen isotope of mass 2 and its concentration (abstract). Phys Rev 39:864

Urey HC, Brickwedde FG, Murphy GM (1932b) A hydrogen isotope of mass 2 and its concentration. Phys Rev 40:1

Urey HC (1947) The thermodynamic properties of isotopic substances. J Chem Soc 1947:562

Urey HC, Lowenstam HA, Epstein S, McKinney CR (1951) Measurement of paleotemperatures and temperatures of the Upper Cretaceous of England, Denmark and the Southeastern United States. Bull Geol Soc Am 62:399–416

Usdowski E, Hoefs J (1993) Oxygen isotope exchange between carbonic acid, bicarbonate, carbonate, and water: a re-examination of the data of McCrea (1950) and an expression for the overall partitioning of oxygen isotopes between the carbonate species and water. Geochim Cosmochim Acta 57:3815–3818

Usdowski E, Michaelis J, Böttcher MB, Hoefs J (1991) Factors for the oxygen isotope equilibrium fractionation between aqueous CO_2, carbonic acid, bicarbonate, carbonate, and water. Z Phys Chem 170:237–249

Valley JW (1986) Stable isotope geochemistry of metamorphic rocks. In: Valley JW, Taylor HP, O'Neil JR (eds) Stable isotopes in high temperature geological processes. Rev Mineral 16:445–489

Valley JW, Graham C (1993) Cryptic grain-scale heterogeneity of oxygen isotope ratios in metamorphic magnetite. Science 259:1729–1733

Valley JW, O'Neil JR (1981) $^{13}C/^{12}C$ exchange between calcite and graphite: a possible thermometer in Greville marbles. Geochim Cosmochim Acta 45:411–419

Valley JW, Bohlen SR, Essene EJ, Lamb W (1990) Metamorphism in the Adirondacks. II. J Petrol 31:555–596

Veizer J (1992) Depositional and diagenetic history of limestones: stable and radiogenic isotopes. In: Clauer N, Chaudhuri S (eds) Isotopic signatures and sedimentary recors. Springer, Berlin Heidelberg New York, pp 13–48 (Lecture notes in earth science, vol 43)

Veizer J (1995) Reply to the comment by LS Land on „Oxygen and carbon isotopic compositions of Ordovician brachiopods: implications for coeval seawater". Geochim Cosmochim Acta 59:2845–2846

Veizer J, Hoefs J (1976) The nature of $^{18}O/^{16}O$ and $^{13}C/^{12}C$ secular trends in sedimentary carbonate rocks. Geochim Cosmochim Acta 40:1387–1395

Veizer J, Holser WT, Wilgus CK (1980) Correlation of $^{13}C/^{12}C$ and $^{34}S/^{32}S$ secular variations. Geochim Cosmochim Acta 44:579–587

Velinsky DJ, Pennock JR, Sharp JH, Cifuentes LA, Fogel ML (1989) Determination of the isotopic composition of ammonium-nitrogen at the natural abundance level from estuarine waters. Mar Chem 26:351–361

Vengosh A, Chivas AR, McCulloch M (1989) Direct determination of boron and chlorine isotope compositions in geological materials by negative thermal-ionization mass spectrometry. Chem Geol 79:333–343

Vengosh A, Chivas AR, McCulloch M, Starinsky A, Kolodny Y. (1991a) Boron isotope geochemistry of Australian salt lakes. Geochim Cosmochim Acta 55:2591–2606

Vengosh A, Starinsky A, Kolodny Y, Chivas AR (1991b) Boron isotope geochemistry as a tracer for the evolution of brines and associated hot springs from the Dead Sea, Israel. Geochim Cosmochim Acta 55:1689–1695

Venneman TW, Smith HS (1992) Stable isotope profile across the orthoamphibole isograde in the Southern Marginal Zone of the Limpopo Belt, S Africa. Precambr Res 55:365–397

Volpe C, Spivack AJ (1994) Stable chlorine isotopic composition of marine aerosol particles in the western Atalantic Ocean. Geophys Res Lett 21: 1161–1164

Wachter EA, Hayes JM (1985) Exchange of oxygen isotopes in carbon dioxide-phosphoric acid systems. Chem Geol 52:365–374

Wada E, Hattori A (1976) Natural abundance of ^{15}N in particulate organic matter in North Pacific Ocean. Geochim Cosmochim Acta 40:249–251

Warren CG (1972) Sulfur isotopes as a clue to the genetic geochemistry of a roll-type uranium deposit. Econ Geol 67:759–767

Watson LL, Hutcheon ID, Epstein S, Stolper EM (1994) Water on Mars: clues from deuterium/hydrogen and water contents of hydrous phases in SNC meteorites. Science 265:86–90

Way K, Fano L, Scott MR, Thew K (1950) Nuclear data. A collection of experimental values of half-lifes, radiation energies, relative isotopic abundances, nuclear moments and cross-sections. Natl Bur Stand US Circ 499

Weber JN, Raup DM (1966a) Fractionation of the stable isotopes of carbon and oxygen in marine calcareous organisms – the Echinoidea. I. Variation of ^{13}C and ^{18}O content within individuals. Geochim Cosmochim Acta 30:681–703

Weber JN, Raup DM (1966b) Fractionation of the stable isotopes of carbon and oxygen in marine calcareous organisms – the Echinoidea. II. Environmental and genetic factors. Geochim Cosmochim Acta 30:705–736

Weber JN (1968) Fractionation of the stable isotopes of carbon and oxygen in calcareous marine invertebrates – the Asteroidea, Ophiuroidea and Crinoidea. Geochim Cosmochim Acta 32:33–70

Wefer G, Berger WH (1991) Isotope paleontology: growth and composition of extant calcareous species. Mar Geol 100:207–248

Welhan JA (1987) Stable isotope hydrology. In: Short course in stable isotope geochemistry of low-temperature fluids. Mineral Assoc Can 13:129–161

Welhan JA (1988) Origins of methane in hydrothermal systems. Chem Geol 71:183–198

White JWC (1989) Stable hydrogen isotope ratios in plants: a review of current theory and some potential applications. In: Rundel PW, Ehleringer JR, Naaagy KA (eds) Stable isotopes in ecological research. Springer, Berlin Heidelberg New York, pp 142–162 (Ecological studies, vol 68)

Whiticar MJ, Faber E, Schoell M (1986) Biogenic methane formation in marine and freshwater environments: CO_2 reduction vs. acetate fermentation – isotopic evidence. Geochim Cosmochim Acta 50:693–709

Whittacker SG, Kyser TK (1990) Effects of sources and diagenesis on the isotopic and chemical composition of carbon and sulfur in Cretaceous shales. Geochim Cosmochim Acta 54:2799–2810

Wickham SM, Taylor HR (1985) Stable isotope evidence for large-scale seawater infiltration in a regional metamorphic terrane; the Trois Seigneurs Massif, Pyrenees, France. Contrib Mineral Petrol 91:122–137

Wickman FE (1952) Variation in the relative abundance of carbon isotopes in plants. Geochim Cosmochim Acta 2:243–254

Wiechert U, Hoefs J (1995) An excimer laser-based microanalytical preparation technique for in-situ oxygen isotope analysis of silicate and oxide minerals. Geochim Cosmochim Acta 59:4093–4101

Wiechert U, Ionov DA, Wedepohl KH (1996) Spinel peridotite xenoliths from the Atsagin-Dush volcano, Dariganga lava plateau, Mongolia: a record of partial melting and cryptic metasomatism in the upper mantle. Contr Mineral Petrol (submitted)

Williams LB, Ferrell RE, Hutcheon I, Bakel AJ, Walsh MM, Krouse HR (1995) Nitrogen isotope geochemistry of organic matter and minerals during diagenesis and hydrocarbon migration. Geochim Cosmochim Acta 59:765–779

Wong WW, Sackett WM (1978) Fractionation of stable carbon isotopes by marine phytoplankton. Geochim Cosmochim Acta 42:1809–1815

Wright I, Grady MM, Pillinger CT (1990) The evolution of atmospheric CO_2 on Mars: the perspective from carbon isotope measurements. J Geophys Res 95:14789–14794

Wyckoff S (1991) Comets: clues to the early history of the solar system. Earth Sci Rev 30:125–174

Xiao YK, Beary ES (1989) High precision isotopic measurement of lithium by thermal ionization mass spectrometry. Int J Mass Spectrum: Ion Processes 94:101–114

Yang J, Epstein S (1983) Interstellar organic matter in meteorites. Geochim Cosmochim Acta 47:2199–2216

Yang J, Epstein S (1984) Relic interstellar grains in Murchison meteorite. Nature 311:544–547

Yapp CJ (1983) Stable hydrogen isotopes in iron oxides – isotope effects associated with the dehydration of a natural goethite. Geochim Cosmochim Acta 47:1277–1287

Yapp CJ (1987) Oxygen and hydrogen isotope variations among goethites (α-FeOOH) and the determination of paleotemperatures. Geochim Cosmochim Acta 51:355–364

Yapp CJ, Epstein S (1982) Reexamination of cellulose carbon-bound hydrogen D measurements and some factors affecting plant-water D/H relationships. Geochim Cosmochim Acta 46:955–965

Yapp CJ, Pedley MD (1985) Stable hydrogen isotopes in iron oxides. II. D/H variation among natural goethites. Geochim Cosmochim Acta 49:487–495

Yapp CJ, Poths H (1992) Ancient atmospheric CO_2 pressures inferred from natural goethites. Nature 355:342–344

Yoshida N, Hattori A, Saino T, Matsuo S, Wada E (1984) $^{15}N/^{14}N$ ratio of dissolved N_2O in the eastern tropical Pacific Ocean. Nature 307:442–444

You CF, Chan LH, Spivack AJ, Gieskes JM (1995) Lithium, boron and their isotopes in sediments and pore waters of Ocean Drilling Program Site 808, Nankai Trough: implications for fluid expulsion in accretionary prisms. Geology 23:37–40

Yuen G, Pecore J, Kerridge J et al. (1990) Carbon isotope fraction in Fischer Tropsch type reactions. Abstr Lunar Planetary Sci Conf. 21:1367–1368

Yuen G, Blair N, Destarais DJ, Chang S (1984) Carbon isotopic composition of individual, low molecular weight hydrocarbons and monocarboxylic acids from Murchison meteorite. Nature 308:252–254

Yung YL, Demone WB, Pinto JP (1991) Isotopic exchange between carbon dioxide and ozone via O (^{1}D) in the stratosphere. Geophys Res Lett 18:13–16

Yurtsever Y (1975) Worldwide survey of stable isotopes in precipitation. Rep Sect Isotope Hydrol IAEA, Nov 1975, 40 pp

Zaback DA, Pratt LM (1992) Isotopic composition and speciation of sulfur in the Miocene Monterey Formation: reevaluation of sulfur reactions during early diagenesis in marine environments. Geochim Cosmochim Acta 56:763–774

Zheng YF (1991) Calculation of oxygen isotope fractionation in metal oxides. Geochim Cosmochim Acta 55:2299–2307

Zheng YF (1993a) Oxygen isotope fractionation in SiO_2 and Al_2SiO_5 polymorphs: effect of crystal structure. Eur J Mineral 5:651–658

Zheng YF (1993b) Calculation of oxygen isotope fractionation in anhydrous silicate minerals. Geochim Cosmochim Acta 57:1079–1091

Zheng YF (1993c) Calculation of oxygen isotope fractionation in hydroxyl-bearing minerals. Earth Planet Sci Lett 120:247–263

Zheng YF, Hoefs J (1993) Carbon and oxygen isotopic covariations in hydrothermal calcites. Theoretical modeling on mixing processes and application to Pb-Zn deposits in the Harz Mountains, Germany. Mineral Depos 28:79–89

Zierenberg RA, Shanks WC, Bischoff JL (1984) Massive sulfide deposit at 21oN, East Pacific Rise: chemical composition, stable isotopes, and phase equilibria. Bull Geol Soc Am 95:922–929

Zobell CE (1958) Ecology of sulfate-reducing bacteria. Prod Monogr 22:12

Subject Index

Springer
and the
environment

At Springer we firmly believe that an
international science publisher has a
special obligation to the environment,
and our corporate policies consistently
reflect this conviction.
We also expect our business partners –
paper mills, printers, packaging
manufacturers, etc. – to commit
themselves to using materials and
production processes that do not harm
the environment. The paper in this
book is made from low- or no-chlorine
pulp and is acid free, in conformance
with international standards for paper
permanency.

 Springer

Printing: Saladruck, Berlin
Binding: Buchbinderei Lüderitz & Bauer, Berlin